D1282488

Soviet Agriculture in Perspective

Other books by Erich Strauss
THE RULING SERVANTS*
EUROPEAN RECKONING
COMMON SENSE ABOUT THE COMMON MARKET*
IRISH NATIONALISM AND BRITISH DEMOCRACY*
SOVIET RUSSIA
SIR WILLIAM PETTY*
BERNARD SHAW: ART AND SOCIALISM

* Also published in the United States of America

Soviet Agriculture in Perspective

A STUDY OF ITS SUCCESSES AND FAILURES

ERICH STRAUSS

And the fourth kingdom shall be strong as
iron . . . and as the toes of the feet were part
of iron, and part of clay, so the kingdom
shall be partly strong, and partly broken.
DANIEL, 2, 40-42

FREDERICK A. PRAEGER, *Publishers*
NEW YORK · WASHINGTON

BOOKS THAT MATTER

Published in the United States of America in 1969
by Frederick A. Praeger, Inc., Publishers
111 Fourth Avenue, New York, N.Y. 10003

© 1969, in London, England,
by George Allen and Unwin, Ltd.

All rights reserved

Library of Congress Catalog Card Number: 69-18402

Printed in Great Britain

Preface

Not more than five years ago, Soviet agriculture was in a state of manifest crisis—a disastrous harvest endangered the food supplies of the population and the feed supplies of herds and flocks. Livestock numbers, and with them the market production of animal products for the consumer, had to be cut back, free-market prices rose and a severe food shortage could only be avoided through massive wheat imports from abroad.

Yet at the end of 1968 the Soviet leaders could look back on a very different scene. Though the season was far from ideal, a near-record grain crop had been safely harvested, most other crops were also quite satisfactory and the supply of meat, milk and eggs was expanding steadily from year to year. This does not mean that, even at official level, the state of agriculture in the Soviet Union received a clean bill of health: capital investment lagged behind the planned figure, the productivity of labour remained low and a number of provinces performed only indifferently well. Nevertheless, there was ample evidence of substantial and sustained progress for all to see. Even those foreign observers who remained suspicious of Soviet statistics could discover for themselves that the Soviet Union was back in the world market as a grain exporter, that its exports of almost one million tons of vegetable oil caused indignant protests by the US Department of Agriculture because they competed at 'unreasonably low prices' with American soya bean in markets as far apart as Western Europe and Japan, and that Soviet exports of butterfat contributed to the difficulties of the 'free' world market in butter.

In 1963-64, Soviet agriculture could fairly be likened to the 'feet of clay' of the Colossus in the prophet Daniel's imagination; by 1968 it seemed to have gained sufficient strength to allow the Soviet Union to engage at least marginally in the acrobatics of the agricultural commodity markets.

The transformation in the character of the main actor was no less spectacular than that of the scene. It is impossible to read the eight volumes of Khrushchev's collected speeches on agriculture—despite the padding, the dreary repetitions and the depressingly flat official 'briefs'—without being struck by their author's sincere feeling for his subject matter. Amongst Communist leaders of the first rank Khrushchev was, indeed, the only one with a genuine interest in, and a considerable knowledge of, agriculture. This was matched by a command of homely and racy language which every one of his listeners could understand, and a lack of ceremony which was probably more refreshing for his audience than for his entourage. Nikita Khrushchev exchanging wisecracks with the peasants of his native village or laying down the law on the best way of feeding cows was one thing, but the First Secretary of the Communist Party publicly criticizing men of almost vice-regal powers in their own bailiwicks and forcing them publicly to admit the error of their ways was a very different matter and did nothing to endear him to the party hierarchy.

It would be difficult to imagine a greater contrast than that between Khrushchev and his successor as leader of the Communist Party. Brezhnev's review of the agricultural situation on October 30, 1968, could find much impressive evidence of progress and soberly reported it, but nobody knew better than the speaker how precarious much of it still was and how much remained to be done. Brezhnev avoided any trace of the 'subjectivism' for which Khrushchev had been blamed so strongly after his fall, but there was a strain of peevish petulance in his exhortations like that of a morose boss giving a general dressing down to his inefficient subordinates.

Agriculture remains perhaps the least predictable and therefore one of the most interesting, if not the most interesting, sector of life within the Soviet Union, though its behaviour is no longer the life-or-death question for the régime which it was during the 1930s and, again, after Stalin's death. After having been for a whole generation primarily a political issue with economic consequences, Soviet agriculture has now turned into a mainly economic problem with political overtones.

One of the main purposes of the present study is to trace this interplay of political and economic forces from the October

Revolution to the historical present. Politically, the dominant question was throughout the relationship between the Soviet power and the mass of the rural population or, for want of a better word, the peasants. Like most political issues, this relationship combined elements of power and elements of exploitation. In simple terms, the basic symbolism of the Soviet emblem —the hammer and sickle—embodies the official ideology of the Soviet system as a union between workers and peasants; what is missing in this somewhat idyllic view is the reality of the Soviet power for whose relationship with the peasants the hammer and anvil would be a more truthful symbol.

Economically, the main issue was originally the place of agriculture in the process of Soviet economic development. The Soviet Union has meanwhile become a great industrial power and this issue is, therefore, of only historical interest for the Soviet Union itself, but not for the struggling peasant countries in other parts of the world for which the success of Soviet industrialization has kindled the hope of a similar short-cut to prosperity for themselves. For the Soviet Union, on the other hand, the main question of current and future agricultural policy is the balance of human and material resources in agriculture compared with other sectors of the economy and the terms of exchange between them.

Politically and economically, agriculture is only one aspect of Soviet society and must be viewed within the wider context of the Soviet system as a whole. To this extent the present study is based on a more comprehensive analysis which I attempted almost thirty years ago in my book *Soviet Russia: Anatomy of a Social History* (1941). Put in a nutshell, it interprets the policies of the Soviet leadership as the response of a ruling bureaucracy to the pressures of social, economic and institutional conditions.

Even at that time the bureaucratic features of the Soviet system were widely known, mainly as a result of the critique of Stalin and Stalinism by Trotsky and his followers. The weakness of this approach was that it regarded bureaucracy mainly as the fault of Stalin and the party machine under his command. Properly understood, bureaucracy was an essential part of the system inherited by the Soviet régime from Tsarism which emancipated itself from the primary social forces of Russian society, monopolized the means of communication between them and thereby subjected them to its own rule.

Since this theory was developed, the study of the Soviet Union has gained immeasurably in width and in depth. Not only has 'Kremlinology' become a recognized, and sometimes illuminating, branch of learning, but there is now—in the West and in the Soviet Union itself—a substantial literary output of genuine scientific value on many aspects of Soviet life. In this literature agriculture occupies a very prominent place, though perhaps less so at the moment than during the crisis period of 1963-64. I have used as much of this literature as was available to me, both from Soviet and from Western sources, but this was by no means all that has been published, and my main aim has been a reasonably coherent assessment of past trends and outstanding problems rather than an encyclopedic deployment of the source material.

The limitations of this attempt are only too obvious to the author and will no doubt become apparent to the reader. They are unfortunately most serious in the chapters dealing with the immediate past and with current problems. However, the student of practical affairs—like the policy-maker—can avoid the mistakes inseparable from basing deductions on inadequate material only at the price of inaction which condemns both to irrelevance.

March 12, 1969   ERICH STRAUSS

Contents

I
Historical Signposts

1917 (November)	Land Decree of the Soviet Government confiscating all land belonging to the landlords and the church immediately and without compensation and putting it under the protection of the local Soviets as a national possession.
1917-18	Division amongst the local peasantry of all landed estates and of most land alienated by peasants from the village community and virtual destruction of all large-scale farming.
1918-21	War Communism with grain monopoly and requisitioning of food supplies for the towns, if necessary by force; growing conflict between the Soviets and the peasants, with general passive resistance and occasional armed clashes.
1921 (March)	Introduction of the New Economic Policy (NEP) and replacement of grain requisitioning by a fixed tax in kind, later changed into an ordinary monetary tax. Gradual removal of restrictions on trade in grain.
1921	Crop failure and widespread famine.
1922 (October)	Land Code protecting the peasants in the possession of land actually cultivated by them, but discouraging the renting of land and the employment of hired labour by larger peasants (*kulaks*).
1923	Scissors Crisis, i.e. sharp deterioration of the peasants' 'terms of trade' with the towns through a growing disproportion between industrial and agricultural prices.
1925 (April)	Relaxation of the legal restrictions on employment of agricultural wage labour and renting of land as part of the policy 'the face towards the village'.
1925	Controversy within the Communist leadership on this policy. Bukharin's slogan *'enrichissez vous'* opposed by Preobrazhenski's principle of 'socialist original accumulation'.
1925-26	Agricultural gross production regains the pre-war level.

13

1925-27   The social differentiation in the village and the exchange between town and country appear as key issues in the struggle between Stalin and his opponents in the Communist Party. The 15th Party Congress decides to link industrialization with the development of large-scale co-operative farming on a voluntary basis.

1928-29   Breakdown of the food supply of the towns on the basis of formally free grain trade. Recourse by the authorities to grain confiscation indicates the abandonment of NEP ('bread front').

1929-32   Compulsory mass collectivization under the slogan of 'liquidation of the *kulaks* as a class' leads to the mass destruction of livestock and other property by the peasants. Development of a network of Machine and Tractor Stations (MTS) as the main instrument of State control over the new collective farms (*kolkhozes*). Creation of huge State farms (*sovkhozes*) in the virgin lands of the south-east as publicly-owned grain factories.

1932-33   Harvest failures and mass starvation lead to a modification of official policy through some reduction in agricultural taxation and the restoration of a partially free market (collective farm market).

1935
(February)   Second Congress of Collective Farmers and Model Collective Farm Statute which grants to the collective farms the use of their land in perpetuity and regularizes the private plots permitted to their members and the numbers of livestock on them. Steep price premia are introduced on deliveries of agricultural raw materials and grain in excess of the normal delivery quotas.

1937   Record harvest and comparative stabilization of the new system.

1938-41   Unfavourable seasons and war preparations are accompanied by tighter controls over the private economy of the peasants.

1941-45   The Second World War causes a sharp decline in agricultural output, partly through enemy occupation and partly through massive losses in manpower and shortages of machinery, fuel and materials.

1946   Severe crop failure.

1947-52	Slow recovery of agricultural production, with the official index barely regaining the 1940 level at the end of the period.
1948 (October)	Announcement of the 'majestic' but abortive Stalin Plan for the Transformation of Nature within ten to twenty years.
1949	Completion of collectivization in the territories gained by the Soviet Union since 1939.
1950	Start of the policy of amalgamation of collective farms and Khrushchev's (unsuccessful) propaganda for the creation of 'agro-towns'.
1952	Growing debate on the place of collective farming in Soviet society. Stalin rejects proposals to transform the collective farms into State farms or to sell the MTS to the collective farms.
1953	Death of Stalin (March) and public revelation of the defects in Soviet agriculture by Khrushchev (September). Announcement of substantial increases in producer prices and reductions in taxation.
1953 (onwards)	Reorganization of the top structure of public agriculture, involving the transfer of local agricultural administration from the Ministry of Agriculture to Party and State authorities at the district level.
1954	Khrushchev launches the 'new lands campaign' for the expansion of grain cultivation in Kazakhstan and Siberia.
1955 (March)	Greater autonomy for State and collective farms in planning production, with detailed regulations from above to be limited to specifying the volume of Government procurements.
1958 (March)	Decision to disband the MTS, to sell most of their plant and to transfer their personnel to the collective farms and to set up Machine Repair Stations.
1958 (June)	Institution of a single price system for Government procurements, with zonal price schedules, instead of the dual prices for quotas and additional sales to the Government.
1958 (autumn)	Record harvest and peak of success of Khrushchev's agricultural policy.
1959	Abolition of the All-Union Ministry of State Farms.
1961	The Ministry of Agriculture is deprived of its main remaining executive functions and limited largely to the organization of agricultural research. Separate

	bodies set up to supply public agriculture with industrial goods and to organize the procurement of agricultural produce.
1962 (March)	The management of public agriculture is reorganized through the creation of Territorial Production Administrations.
1962 (June)	Sharp increases in the prices of livestock products at producer and retail level.
1962 (November)	Khrushchev obtains against considerable opposition a division at local level of the Communist Party organization into an industrial and an agricultural branch.
1963	Serious crop failure followed by a sharp contraction in livestock and the purchase of 12 million tons of wheat in America.
1964 (July)	Khrushchev announces old age pensions and other social security benefits for collective farmers from January 1965, with State contributions to their cost.
1964 (autumn)	Khrushchev's resignation, immediately followed by restoration of the unitary organization of the Communist Party.
1965 (January)	The executive functions of the Ministry of Agriculture are restored and V. V. Matskevich is reappointed Minister.
1965 (March)	Brezhnev announces substantial increases in producer prices, restores premium prices for extra deliveries, reduces taxes on agriculture and limits the quotas for deliveries to the State. Price restrictions in collective farm markets abolished and plans to build better markets announced.
1965 (August)	Unfavourable harvest and negotiations of large-scale wheat purchases from Canada, Argentina and other Western countries.
1965 (December)	The Budget for 1966 contains a substantial reduction in the collective farm tax; the rural addition to retail prices to be abolished and prices of farm machinery to be reduced.
1966 (January)	The Communist Party announces the decision to convene the Third Congress of Collective Farmers and sets up a commission under the chairmanship of L. Brezhnev to work out a new charter to replace the one drawn up in 1935.
1966	Draft Five Year Plan for 1966-70 adopted by the

16

(March/April) Twenty-third Congress of the Communist Party, providing for record investment in agriculture and modest output goals.

1966
(July)

Decree introducing guaranteed monthly wage for collective farmers at rates corresponding to the State farm norms, to be paid twice monthly and to be supplemented by special bonuses at the year end.

1966
(autumn)

Record harvest, particularly of cereals, and complaints about the insufficient capacity of the food processing industry to cope with rising supplies of agricultural produce.

1967
(spring)

Pilot scheme to be initiated on 390 State farms, introducing an interest charge on capital and authorizing the use of a high proportion of planned profits for incentive fund and other types of local use.

PART ONE
Starting Points

II

The Permanent Challenge
of Soviet Agriculture

The Main Issues
The comparative failure of Soviet Russia on what was once called, without exaggeration, the 'agrarian front' has not only been accepted as a fact by most outside observers but also, by implication, by the Soviet authorities themselves. Its symptoms are a low volume of agricultural production per head of the population, generally accompanied by high costs of production, periodic food difficulties, either in the quantity or quality of supplies or both, and limited supplies of raw materials for such important processing industries as the food industry, textiles and leather goods. The sometimes obsessive preoccupation of the Government with agricultural policy and administration provides evidence of the stubborn persistence of unsolved problems and of their critical political importance.

Agriculture and the relations between the Soviet power and the peasants were among the determining factors at three seminal turning points of Soviet economic policy—the introduction of the NEP in 1921, the great industrialization drive connected with the practice of embodying economic objectives in Five Year Plans in 1928, and the 'thaw' after Stalin's death in 1953. Each of these is inevitably linked with the name of the Party leader who was representative of the period and in authority at the time of the most important decisions—Lenin, Stalin and Khrushchev.

If the NEP had become inevitable at the end of the Civil War period, it was historically no more than a breathing spell. It was accepted, genuinely enough, as the economic framework for the co-existence of the post-revolutionary village, which contained more than four-fifths of the population, and of a Communist dictatorship in the towns which had temporarily lost most of their productive resources. Nevertheless, the consequences of this policy demonstrated within a few years that it was virtually impossible to contain the development of Soviet society within its limits for any length of time.

Whatever may be thought of the choices open to the Communists when a final impasse was reached in 1928, few people will now regard *compulsory* collectivization as anything but a major social and economic catastrophe. The man-made famine in which it culminated was apparently described by Stalin himself as the ultimate horror for its far from squeamish author, exceeding even the gravest emergencies during the war with Nazi Germany.[1] In contrast with the struggle against the Germans, the peasant war of 1929-33 did not end in a clear-cut victory for the Soviet Government; on the contrary, at the time of Stalin's departure from the scene in 1953 the situation contained again all the elements of a new agricultural crisis which threatened to be no less serious for being chronic rather than acute.

The problem of agriculture was the theme-song of the Khrushchev era from beginning to end. The new leader himself had been one of the foremost agricultural experts in the Soviet hierarchy ever since the war, and he identified himself completely with the agrarian reform policy which he introduced personally within six months of Stalin's death in a famous, and brutally frank, disclosure of the true position in the countryside.

Khrushchev's rise to supremacy, though never to absolute power, was linked in time and substance with the partial success of his policy of raising the level of agricultural output through a gigantic crash programme designed to get quick results at the cost of taking considerable risks. The revival of the threat of agricultural stagnation after 1958 may well have been one of the causes of the increasingly erratic policy of the Soviet Government under his leadership in many fields. The feverish, and almost desperate, attempts to improve the position of agriculture through repeated administrative reshuffles and the severe setback of the crop failure of 1963 must have undermined his position in the party hierarchy and contributed to his downfall in October 1964.

The close interaction between economic and political factors in the agrarian policy of the Soviet leadership is evident at every important stage. However, it would be equally futile to regard this policy essentially either as a power struggle pure and simple or as a large-scale essay in agricultural economics, to be marked academically like an outsize examination paper. In order to be of any value, a critical assessment must show the relative contributions of socio-political and economic factors to the final outcome at least in principle, and ideally in their full complexity. This involves inevitably an historical survey, but such an interpretation of the process as a whole may benefit from a preliminary review of the underlying problems, some of which are peculiar to Communist Russia, while others may depend only partly on the national background.

In the first place, an attempt will be made to outline the special part played by agricultural problems in the internal evolution of the Soviet régime and to assess the consequences of its international position. The following chapter will be devoted to a short analysis of the bureaucratic character of the Soviet régime and to the effects of this fact on agricultural policy. The preliminary analysis will be completed by a discussion of the place of agriculture in economic development, both in the early stages of industrialization and after, against the Russian background.

The Basic Paradox of the Soviet Régime

Almost from the very start of a socialist movement in Russia, its adherents were at loggerheads about the future consequences of the social and economic backwardness of the country. This issue was already implicit in the problem put to Karl Marx by Vera Zassulich and others, as to whether the ancient village community would have to disappear or whether, on the contrary, it could become the nucleus of a socialist society.[2]

Russian socialists were united in their root-and-branch opposition to Tsarism, the landlords and the growing bourgeoisie of the towns, but they were deeply divided in their views about the system which should take the place of the *status quo* after a successful revolution. Three main groups could claim to be considered, at least potentially, as genuine political parties, the

23

Narodniki, or Social Revolutionaries, the Mensheviks and the Bolsheviks.

Though most *Narodniki* professed a vague kind of socialism, they looked to the peasants as the backbone of the new Russia. The overthrow of Tsarism and the landlords was for them an essential step towards the complete liberation of the peasants from the remnants of feudalism which had survived the half-hearted abolition of serfdom in 1861. They were essentially a party of the peasantry, with a potentially powerful appeal to the majority of the population which was enhanced rather than reduced by their division into a right wing and a left wing according to the revolutionary lengths to which they were prepared to go to attain their aims.

The other two socialist groups regarded themselves first as members and later on as 'fractions' of the Russian Social Democratic Party because of their common acceptance of Marxist theory and their belief in the industrial working class as the main agent of the coming revolutionary struggle. Otherwise they had little in common, just as the 'Social Democratic Party' itself had no independent existence after the full implications of the split between the two groups had become clear.

The *Mensheviks* were rightly impressed with the enormous difficulty of establishing a socialist system of society in a country where the industrial working class was a small minority and where large-scale industry was the exception rather than the rule. From this they concluded that in the post-revolutionary society the capitalists would have to assume the economic leadership of the country and thereby effectively disqualified themselves from running the revolution.

The *Bolsheviks,* or as they called themselves from 1918 onwards the *Communists,* were marked out as the future rulers of the country by their insistence on the primacy of grasping power as soon as an opportunity offered, but the 'contradictions' of a socialist government in control of a backward peasant country, which had once haunted the endless discussions of the tiny groups of conspirators, returned after the revolution in broad daylight to bedevil the policies of the first 'Workers' State' in history.

For centuries, the relationship between the Russian Government and the peasants had been essentially simple: The Tsar himself may have been regarded as a symbol of an ultimately just order, but the State was a moloch which had to be fed continuously with taxes and periodically with the lives of young recruits, it was an enemy to be avoided, if possible, to be endured, if it could not be avoided, and to be fought, if it became unendurable, though the price of rebellion was invariably large-scale slaughter. At first the Communist State seemed to differ from its Tsarist precursor by applauding the great peasant risings at the end of the First World War instead of trying to put them down by force—and the peasants, or at least many of them, regarded it as their ally against the forces of counter-revolution and intervention which were only too plainly tarred with the old brush of oppression and exploitation.

But the peasants' age-old dream of owning the land instead of being attached to it and of doing what they liked with its produce proved no less impossible of realization under the new dispensation than it had been under the old one. The new rulers applauded the revolution which had driven out the landlords, but in the emergency created by the Civil War they were just as hungry for men and taxes as their predecessors, though they took the product of the peasants' labour in kind rather than through the tax gatherer; and even after the acute military crisis was over, and taxes again replaced armed requisition, the peasants' hopes of being left alone to do as they pleased were soon disappointed.

The new régime was essentially urban, and neither the Communist Party nor the machinery of the State was particularly strong or effective in the countryside. Its mass basis was virtually confined to the towns, and particularly to the industrial workers, whom the end of the emergency found in a sorry condition. The years of fighting had consumed all current material resources and reduced needs to bedrock—above all to the daily bread for which the urban population was entirely dependent on the peasants, to whom it had to offer very little in exchange.

The weakness of Russian capitalism had permitted the Communists to gain absolute power relatively easily, while the Labour movement of the advanced countries had to be satisfied, at best, with the place of junior partner in the management of the established social order: now, in retrospect, this strategic political boon turned out to have grave economic disadvantages for the victors. To this extent the Mensheviks were proved right after the event, for the lack of a solid industrial heritage deprived the Communists of the essential material resources for setting up the socialist society to which they were sincerely dedicated. They were, in fact, reduced to the unglamorous position of having to drive the hardest possible bargain with the *muzhiks* in order to keep alive and to obtain sufficient raw materials to start the wheels of industry turning again. As most of the peasants were themselves hovering on the brink of subsistence, this was not a pleasant situation for a socialist government to be in, and even when conditions improved somewhat after a few years of peace and economic recovery, they were at best not better than they had been at the time of the outbreak of the First World War.

The 'workers' and peasants' republic' was thus a very different place from the expectations of its creators and the same was even more true of its next stage, which took the form of a grim duel between the Soviet power and the peasants. The Government's original plan to split its opponents by the use of 'social engineering' was, on the whole, a costly failure. The division of the village between *kulaks,* middle peasants and poor peasants, though real enough in its way, was not nearly so deep as the gulf of conflicting interests between the peasantry as a whole and the towns as represented by the political rulers of the country. The culmination of this mistaken official policy in the 'liquidation of the *kulaks* as a class' inflicted a setback on agricultural production from which it had barely recovered by the Second World War.

International Isolation and Soviet Agriculture
Until the defeat of Nazi Germany, the Soviet Union was the only country ruled by professed enemies of the capitalist system. For most of the time, it lived in a state of acute tension with some powerful industrial nations, even when there were no open hostilities. Defence needs thus played a prominent part in the economic policy of the Soviets and absorbed a growing proportion of the inadequate resources of the country—as events were

to prove, with full justification.

However, the international isolation of the Soviet Union also had economic consequences in the narrower sense of the term. The distrust of foreign investors in the future economic policy of the Soviet Government, following the repudiation of all Tsarist debts, condemned the half-hearted official attempts to attract foreign capital through 'concessions' to virtually complete failure. The Soviets proved extremely punctilious in honouring their own financial obligations abroad, and therefore enjoyed good commercial credit on current account, but the large capital investment needed for the industrialization programme had to come from domestic sources. Initially this meant, above all, from the peasants.

The necessity to industrialize the country in the shortest possible time was as much political as economic and demanded as much by international as by domestic considerations; hence the overwhelming emphasis on heavy industry with its special relevance for modern armaments. The need to industrialize entirely from domestic resources involved a task reminiscent of the famous feat of Baron Muenchhausen, who boasted of having pulled himself by his hair out of a bog in which he was floundering. The pre-1914 industry of Russia, such as it was, had been largely financed from abroad, and foreign capital played an apparently indispensable part in the transition from a backward subsistence economy to modern industrial development. Foreign assistance was, of course, also important in the reconstruction of the Soviet economy after 1928, but it was available only in the form of imports of machinery and technical advice which had to be paid for practically at once. The foreign exchange needed for this purpose had to come from exports of grain and other agricultural products, even though these had to be extracted from the peasants by force and at incalculable cost in human suffering and economic efficiency, and even though the population had to go short of food.

The 'original socialist accumulation' analysed in advance by Preobrazhenski took practical forms which deserved only too well the famous words of Karl Marx about the 'so-called original accumulation' that 'capital was born dripping blood and mud from top to toe and from every pore'.[3] In the present

27

context the point to remember is the close connection between the international isolation of the Soviet Union and the fact that industrialization took place within the framework of a veritable war between the Government and the peasants.

However substantial the involuntary share of the peasants in the financing of capital investment in the first place, it was not in the nature of a once-for-all contribution extracted from them under the spur of necessity and after which they were allowed to go their own way. Large-scale industrialization involves a mass movement of population from the countryside to the towns and the replacement of food producers by food consumers. The resulting growth of the market for agricultural produce could be of great advantage to the remaining agricultural population, just as the greater output per head of the new industrial workers compared with their often very small output as peasants is normally of great benefit to the national economy. Frequently, though not invariably, this process has involved the transformation of a country exporting food and agricultural raw materials into a net food importer, if the pace of the changeover from agriculture to industry outstrips the increase in agricultural productivity.

The international position of the Soviet Union was—and is —such as to exclude the reliance on food imports as a normal economic occurrence, because it stands outside the system of international division of labour which would have made such a transformation possible. Quite apart from the question how far Soviet industrial products would be competitive in international markets and could raise the foreign exchange needed to finance large-scale regular food imports from capitalist countries, the political risks of import-dependence in such a vital sphere would obviously, and reasonably, be regarded as intolerable by the Soviet Government. The emergence of some kind of community of Communist countries after 1945 may have created some possibilities of international division of labour within the Soviet bloc, but not only are its members busily developing their industries, they generally look to the Soviet Union for certain agricultural produce in exchange for their own manufactures.

In such circumstances the expansion of agricultural output occupies a key position in economic development strategy. As at the same time population moves from farm to factory, agricultural productivity must grow fast enough to compensate for the fall in man-power and to provide, in addition, a larger absolute volume of output from fewer workers. This is the sense in which the problem of agriculture forms a permanent challenge to the Soviet leadership.

In retrospect it is clear that the uneasy relations between the Soviet power and the peasants have been the most formidable constraint on its internal freedom of action on a number of occasions. However, within a closed economy the link between the growth of industry and the ability of agriculture to supply the expanding population with food and the expanding traditional consumption goods industries with basic raw materials is of more than historical interest. Until and unless the agricultural basis of the system is secure, the system as a whole remains under pressure. It would be idle to deny that the Soviet Union has achieved considerable successes in the agricultural field— everybody knows about the disastrous reverses it has suffered from time to time—but permanent success has so far eluded it. Why this should be so, and what are the conditions for achieving such success, is the subject of the following analysis.

III

Bureaucratic Dictatorship and Agricultural Policy

Main Features of a Ruling Bureaucracy
The internal social structure of the Soviet Union and its international position were, in the final analysis, the main factors which determined the contents of Soviet agricultural policy, as of so much else; the bureaucratic form of Soviet life exercised an almost equally complete and decisive influence on the methods employed in carrying this policy into effect.

A detailed analysis of the origin and causes of Soviet bureaucracy cannot be given in this place,[1] but a short recital of its main characteristics and of their relevance to agricultural policy is needed in order to understand past problems and the present situation.

The Soviet system of government is the heir of the Tsarist bureaucracy which ruled Russia from about the sixteenth century onwards and which received its classical shape in the reign of Peter the Great at the beginning of the eighteenth century. Bureaucratic government in Russia is, therefore, a deep-rooted tradition which has survived radical changes in social conditions, though not without involving itself, and the whole country, from time to time in disastrous difficulties. It is a 'ruling bureaucracy'[2] in the sense of a power élite largely independent of primary social forces, and exhibits the characteristic features of such a system in almost classical completeness. Though the mere effluxion of time has proved its fitness for survival, history has demonstrated at the same time its unfitness for shaping the policies of the large empire over which it has ruled in the manner most appropriate to the circumstances of the time.

Put in a nutshell, bureaucratic rule has been largely responsible for the fact that Russian history during the last one-and-a-half centuries has progressed through a whole series of historical vicious circles on a gigantic scale. These cyclical movements started as a rule with a period of virtually unchallengeable bureaucratic absolutism which appeared no doubt to its beneficiaries as the normal state of affairs. However, sooner or later it led to a grave public calamity which made the maintenance of the *status quo* impossible and discredited its adherents. This stage was followed by a period of more or less radical reforms which temporarily improved the situation, but which sooner or later ran into successful opposition by the entrenched bureaucracy; from there it was only a single step to the reassertion of bureaucratic absolutism which in due course again ended with a public calamity, if possible on an even larger scale than the previous time. The culmination of this process under the old régime was the overthrow of Tsarism by the February Revolution of 1917 and the victory of the Communists in November of the same year.

The new rulers effected changes in the bureaucratic structure of government which were, on the surface, extremely radical. The most reactionary régime in Europe was succeeded almost overnight by a government of ardent revolutionaries whose leaders spurned even the title of Ministers and proudly described themselves as 'People's Commissars'. In the higher echelons of the bureaucracy there was an almost complete change of personnel and most of the new leading officials were recruited from outside the ranks of the old hide-bound bureaucracy. This infusion of new blood was accompanied by a change of spirit and ethos, with a violent reaction against the oppression, sloth and corruption of the old régime.

In the short run, this revolutionary upheaval virtually paralyzed the government machine which had already been grinding to a halt during the last stages of the First World War, and particularly since the February Revolution. The political and administrative vacuum created in this way was filled by the Communist Party rather than by the haphazard institutions of the new Soviet State. Under the almost unbearable strain of the years of War Communism, the Communist Party itself tended to assume the character of a ruling bureaucracy through which the leaders of the Party exercised a dictatorship over the whole people limited only by its lack of effective power in certain fields.

The most perceptive early critics of the Leninist conception of the political process, such as Leo Trotsky and Rosa Luxemburg, had pointed out long ago that this development was potentially inherent in the structure and ideology of the tiny Bolshevik 'fraction' of the Russian Social Democratic Party, who 'represented in their closely knit and ruthlessly led secret organization the nucleus of an anti-State . . . constructed in the negative image of the system'[3] which they wanted to overthrow. The transformation of the Communist Party into the effective government of the country therefore took place surprisingly quickly, and with the forcible suppression of all other political movements it soon combined a monopoly of political activity with the supervision and control of the government machine.

Despite the cataclysmic changes undergone by this machine after the Soviet revolution, it remained, in Lenin's telling phrase, the old Tsarist system only thinly anointed with Soviet holy oil,[4]* and its size and bureaucratic proclivities increased with the enormous growth of government activities in the economic sphere. Conventional inspection arrangements, even with the hopeful title of 'Workers' and Peasants' Inspection', proved no less futile than they had done in the Tsarist past, and the control of the operations of the State machine by the political leadership was frequently frustrated by the persistence of bureaucratic rivalries and defects.

The most spectacular and sinister development was the emergence of the Secret Police as an independent power during the later stages of Stalin's rule: at times it threatened to dominate all other groups, not excluding the Communist Party machine. The coalition of its frightened opponents after Stalin's death was strong enough to eliminate Beria, the chief of the Secret Police, and to reduce its status. This threat of complete police rule may have shocked the whole establishment sufficiently to curb the powers of the police for good, though the conditions

---

* Forty-five years later, N. S. Khrushchev was to complain that the new Territorial Production Administrations were functioning in the old way, because their personnel had brought 'the old luggage with them' from their previous posts.[5]

which made this threat possible in the first place need not have disappeared permanently. It is at least suggestive that after Khrushchev's fall one of Beria's successors, Shelepin, graduated from this post to the Party Praesidium at the unusually early age of 46, though this does not seem to have enabled him to contest the advancement of Brezhnev to the leading place in the party hierarchy.

The bureaucratic dictatorship in control of the Soviet Government is in important respects similar to the Tsarist bureaucracy, but its strength as well as its weaknesses are exaggerated by the fact that its size and the scale of its operations are incomparably greater:

(i) In the first place, there is the enormous concentration of power at the top of the bureaucratic machine. This consists of two pillars of unequal political strength: the Party and the Government. The Party is the ultimate repository of power, the source of energy and the seat of control; the Government is the agent in charge of day-by-day administration. In emergencies, such as a change in leadership and the selection of a new team, the supremacy of the Party asserts itself. In the running of ordinary affairs, the top leadership of the Party and the Government coalesce, but the machines operate separately. Khrushchev's attempt to involve the Party administration directly in the running of agriculture was a symptom of the seriousness of the agricultural situation rather than a precedent to follow.

(ii) The prohibition of any independent centre of social, political or economic interests outside the recognized official bodies limits the expression of such interests to their influence on the sections of the Party and the Government officially in charge of their respective spheres, though during the later 1920s the economic advantages of party membership made rural Communists the natural allies of the better-off sections of the peasantry.[6] Clashes of interest between social groups thus tend to assume the form of difficulties in the operation of the bureaucratic machine, and particularly of friction between different bureaucratic organs. The function of the bureaucracy as intermediary between different interests tends to disguise these interests to such an extent that not even the top leadership may be aware of the true causes of so-called administrative problems.

B

(iii) Excessive centralization distorts the system of upward communication of information and impedes advance notice about important developments, particularly if they are not in line with current policy, until they cause administrative difficulties; frequently the ignorance of the local agents of the bureaucracy about the true situation makes their reports incomplete or downright misleading.

(iv) With policy determined at the top on the basis of defective, and often out-of-date, information, adjustment to important external changes will also be late and incomplete. As a rule, the machine continues to operate on the basis of existing policies until it comes up against obstacles which do not yield to local pressure. This may cause more or less prolonged deadlock, until a new policy is agreed at the top and communicated to the lower echelons. The whole cumbersome machine thus tends to operate through changes of direction rather than of gear.

(v) As bureaucratically determined policies essentially take the form of instructions about the behaviour of bureaucratic organs, minor changes in policy are apt to take the form of dismissals and replacements of individual officials who have failed to execute such policies to the satisfaction of their superiors; major policy changes often involve reorganizations of the bureaucratic machine itself. As major administrative difficulties are often caused by unsolved social problems, the rulers may well mistake such reorganizations as genuine solutions of the underlying problems.

(vi) As long as an official policy remains in operation, all events will tend to be interpreted in line with it at all levels of the bureaucratic pyramid. Given the bureaucratization of the whole of public life, virtually all public utterances will support the ruling policy and condemn all facts and opinions which do not conform to it. Conversely, a radical change in official policy involves everybody—administrators, propagandists and even scholars—in a corresponding 'revaluation of all values' compelling them to condemn what they had preached before and to preach what they had condemned before.

(vii) These characteristic defects of a ruling bureaucracy extend to the organization and control of economic policy. The whole process of forced industrialization took place within this framework, and economic planning was developed on thoroughly bureaucratic lines. Methods of decision were excessively centralized and local management was subjected to specific direction from above even in matters of detail. The obvious defects of such a system produced from time to time large-scale reorganizations, but they never extended to the fundamentals of the bureaucratic power structure and were, therefore, invariably unsuccessful. Confronted with such a caricature of economic planning, most Western critics have found little difficulty in claiming that a planned economy is inherently inferior to a market economy, though the ideal model of the latter frequently implied in such comparisons bears as little relationship to the realities of Western capitalism in the last third of the twentieth century as the activities of the economic planning branch of the Soviet economy to the theoretical needs of a socialist economy.

The effects of the Soviet system of bureaucratic rule extend to all the main aspects of its agricultural policy, the social and political relations between the Government and the rural population and the organization of production and marketing of agricultural produce.

The Bureaucracy and the Peasants

During the first ten years of Soviet rule, the peasants stood socially and economically outside the Communist system. They paid their taxes, if they could not avoid them, sold their surplus production in the market if and when it seemed advantageous to them and bought the few manufactured goods within their reach at the best possible prices, which were generally quite high. At the same time, they were subject to the political consequences of life under a government which pursued aims radically differing from their own and which was capable of enforcing its will within certain limits. Even at that early time the relations between the régime and the peasants were influenced at critical moments by the fact that the Government was a political dictatorship organized in a bureaucratic manner.

The Government represented a concentration of power employing a centralized administrative machine, while the peasants were a completely atomized mass of some four and twenty millions of small households. This situation not unnaturally created an illusion of administrative omnipotence on the part of the authorities which was not fully justified by the underlying balance of social forces. As a result, the policy of the Soviets towards the peasants was frequently governed by a 'law of over-success',[7] i.e, by the apparently complete enforcement of official policy followed by dangerous, but at first imperceptible, reactions on the part of the peasants. Friction between the régime and the peasants was inherent in the conditions of the time, but the form it took and its destructive violence were in no small measure caused by the bureaucratic character of the régime.

Once the decision had been taken to link industrialization with the creation of producers' co-operatives in the villages, the bureaucratic machine took over with devastating results. The most ambitious objective of the First Five Year Plan was to collectivize about one-seventh of the total arable area by the end of the period,[8] but once the full resources of Party and State had been brought into action, this objective was 'achieved' in a few months, and collectivization accompanied by the 'liquidation of the *kulaks* as a class' swept the country with fire and sword. (A convenient illustration of the administrative consequences of the new policy was the number of laws and regulations which it entailed. The *History of Collective Farm Laws* lists 92 All-Union laws and 26 for the Russian Federated Socialist Republic until 1928, a total of 118. This rose to 225 or about 1 per week for the next four years, and a further 265 laws and regulations were enacted during 1933-36, the years of final crisis and compromise.[9])

Mass collectivization extended the administrative power of the bureaucracy over the village, mainly through the medium of the Machine and Tractor Stations (MTS) and the effective subordination of the collective farms to the local representatives of the Party and the Government. However, by 1933 the number of collective farms was well over 200,000 and continued to increase until the war. The administrative difficulties of coping with such large numbers were, of course, extremely great, though less unmanageable than the control of some twenty-four million individual peasants would have been.

A few years after the end of the Second World War, the authorities decided to reduce the numbers and increase the size of collective farms, and at the end of the first year of a determined drive, in 1950, their number had been approximately halved; three years later it had fallen to only 93,000 and at the end of 1964 there were only 37,600 collective farms left.[10] The economic arguments behind this massive increase in the size of the main operative unit were complex, but its main driving force was almost certainly administrative.

As the grip of the bureaucracy over the village tightened, the temptation to mistake social friction and clash of economic interests for administrative errors became ever more compelling. Under Stalin, the relationship between the state power and the peasants had been frozen in a rigid pattern of external pressure and control which lasted almost unchanged for twenty years; under Khrushchev it became both closer and more problematical and, given the nature of Soviet government, the difficulties of this relationship were inevitably interpreted in administrative terms. Thus the familiar ding-dong game of centralization, followed by decentralization and recentralization, which had become a normal feature of industrial management at a very early stage, now extended to the countryside.

Khrushchev's famous Report to the meeting of the Central Committee of the Communist Party in September 1953 contained an uncompromising attack on the Ministry of Agriculture as an unwieldy bureaucracy, remote from the object of its operations and suffering from delusions of grandeur: 'The planned tasks for the collective farms in agriculture and animal husbandry list over 200 to 250 goals . . . an excessive number of tasks has brought about an extraordinary inflation in every type of report-making . . . compared with pre-war, the accounting indices of the collective farms have increased almost eightfold.'[11] This was not simply the rhetorical flourish of an ebullient orator but the opening shot in a determined power struggle which produced some strikingly plain speaking, including Khrushchev's demonstration that the structure of the Soviet Ministry of Agriculture differed very little from that of the Tsarist Ministry—except in the larger size of the Soviet bureaucracy.[12] 'The October Revolution took place in the country, a revolution took place in agriculture, we replaced the small peasant economy by a large-scale socialist economy, but the structure and machinery of management of the *ancien régime* remained essentially unchanged."[13]

Khrushchev's early reforms included some determined steps in the direction of increasing the independence of the men on the spot, but after some years of hesitation the pendulum began to swing back, until the establishment in 1962 of a new 'complex Gargantuan pyramid characterized by a dual system of control and administration'[14] in the shape of the Territorial Production Administrations. Some months later Khrushchev drew the ultimate—and probably suicidal—conclusion from this policy of integrating agriculture into the bureaucracy by splitting the local party organization itself into an industrial and an agricultural branch, probably with fatal consequences for his personal position.

However, the practical effect of the new system for which Khrushchev risked so much remained disappointing: a few months before his fall, in April 1964, he complained in a memorandum to the Party Praesidium that the TPA did exactly the same as all the other bodies had done, because they were composed of the very men who had been in charge of Soviet agriculture for the last twenty years, with the same habits and the same style of work.

The circle was completed within a few months of Khrushchev's fall from power by the restoration of the administrative powers of the Ministry of Agriculture under the same chief whom Khrushchev had demoted a few years earlier.

Bureaucracy and Agricultural Production

'In a very real sense, exhortation, political activity, administrative reorganization and a tightening of central controls is substituted for more realistic production targets and greater allocation of resources to agriculture.'[15] There is, in fact, a close connection between the gradient of the agricultural output curve and the temper of the Soviet bureaucracy, but the bureaucratic system of government influences farm production policy by more direct methods.

Just as in the First Five Year Plan the original collectivization

38

aims of the State Planning Commission were pushed aside as too moderate and replaced by mass collectivization, the original plans for expanding state farming were also superseded as pusillanimous. Instead, giant units with tens of thousands of hectares were set up with complete disregard of costs and without the necessary expert knowledge of managing such huge enterprises —if expert knowledge would have sanctioned their being set up in the first place. This enormous misdirection of effort was another manifestation of bureaucratic absolutism at work in an environment which could not offer effective resistance to the plans of the rulers.

Western observers frequently criticize the Soviet tendency towards sweeping 'campaigns' which apply essentially sound ideas of limited practical use in an exaggerated manner in both suitable and unsuitable conditions and frequently do more harm than good.[16] This was true of the rotational grassland policy connected with the name of V. R. Vilyams (Williams) under Stalin as much as of Khrushchev's favourite panacea of growing maize in dry and cold regions. Khrushchev himself argued at first that he was not attacking the grassland system as such but only its 'routine application'[17] and inveighed against the domination of routine in agricultural policy; soon afterwards he proceeded to attack the system root and branch in another 'campaign' which had all the defects of the system he criticized.

This tendency is, in fact, essentially independent of the individual judgment and temper of the man at the top of the bureaucratic machine and is a feature of bureaucratic management as such. The mere fact that the acknowledged leader expresses in public an inclination towards a certain policy at the expense of another sets up a strong current throughout the whole pyramid. Nobody can afford to remain identified with a policy out of favour in high places, everybody wants to show his eagerness to please his superiors by applying the approved policy within his own bailiwick and refuses to be deterred from showing his zeal by unfavourable conditions—rather the contrary.

Nor is this exaggerated conformism confined to practical men of affairs: writers, economists or scientists who advocate, however sincerely, a discredited or merely unfashionable line of

39

action find themselves under a cloud, unless they show exceptional mental and moral agility. Conversely, the spokesmen for the currently fashionable policy exaggerate its successes and the less scrupulous amongst them make the figures fit their claims where they do not do this naturally. The discovery of the real meaning behind subtle shades of expression, the selection of some feature for comment and the neglect of others or slight shifts in statistical presentation are a fertile hunting ground for experienced Kremlinologists. Agricultural statistics are a particularly rich quarry for such investigations, because the frequent failures in this field constitute a standing temptation for official spokesmen to put their best foot forward at the expense of accuracy and objectivity.

One of the easiest methods of attaining statistical excellence on paper is the use of an 'anti-telescopic bias'[18] in the yardsticks employed; this is done by writing down the results of a past period which are then used as a basis for comparison with current results which may, of course, also be overstated. Such methods are more difficult to apply to physical quantities than to aggregates and indices, and the critique of the official output indices has been developed almost to a fine art by Western analysts. However, the blatant liberties taken with grain production figures since 1933 show that even physical data can be manipulated in case of need.[19]

The practices adopted by the Soviet power in its treatment of agriculture show any number of cases where agricultural policy has been affected by the bureaucratic character of the régime. Khrushchev's exposure of the true state of agriculture at the time of Stalin's death abounds with illustrations. The practice of counting livestock numbers at the beginning of the year, perfectly harmless in itself, viciously distorted the seasonal flow of meat production, because animals were not slaughtered until after the magical date of January 1st; this enabled managers and party officials to show larger increases in livestock, whatever the effects on fodder usage and meat supplies.[20] The distribution of official programmes of production and procurements was a permanent source of divergence between rational use of resources and administrative convenience: 'If a district is given the task of planting 100 hectares of cabbages, it invariably distributes the task among all the collective farms. The same is true of cucumbers, tomatoes and other crops. Economically this principle is completely incorrect. . . . We give the same task to a collective farm with well-watered soil and close to water as to a collective farm without such lands. Collective farms are forced to plant cabbages and other vegetables on unsuitable land and gather low harvests with much labour and material waste.'[21]

A clear understanding of the nature of the Russian system of government, of the obvious strength as well as the inherent weaknesses of a ruling bureaucracy, is thus essential for a realistic analysis of Soviet agriculture, as of other areas of life in the Soviet Union.

IV

Agriculture in
Soviet Economic Development

1. AGRICULTURE AND ECONOMIC GROWTH

Although the most backward of the European Great Powers, Russia was in 1914 well on the way towards having the basic forms of manufacturing industry. Its industrial system was, however, heavily biased towards the needs of the army and towards simple consumer goods which accounted for almost three-fifths of total production.[1] War, revolution and civil war caused a cruel setback, but by about 1926 the two main branches of the national economy, agriculture and industry, had approximately regained their pre-war level of output. According to the fashionable theory, Russia was supposed to have completed the stage of 'economic take-off' to self-propelled further growth at the time of the First World War;[2] in fact, conditions towards the end of the NEP period gave little promise of semi-automatic progress.

In the late 1920s, as before the First World War, only 18 per cent of the population of the enormous country lived in the towns, with the rural population, which was predominantly employed in agriculture, accounting for 82 per cent of the total. In this respect the Soviet Union was similar to other backward agricultural nations trying to modernize their economy through rapid industrialization. It is true, it had the advantage of a comparatively broad, if old-fashioned, industrial base; on the other hand, the international isolation of the régime precluded recourse to overseas capital in any form, whether through direct investment, long-term loans or 'aid'. Other developing countries are discovering through painful experience that agricultural development is an essential condition of sustained economic growth, even when they have access to foreign capital, but in the Soviet Union domestic agriculture was from the start the main

42

source of the material as well as of the human resources for forced investment in industry. This fact made the relationship between the Soviet power and the peasants the focal point of a violent struggle for the mobilization of these resources, and the nature of the economic links between agriculture and the rest of the economy assumed a similarly critical position in the economic policy discussions of the time.

Agricultural Gross Production and Market Production

The simplest, and easily the most important, reflection of the economic realities of Russia's post-revolutionary peasant agriculture was the 'grain balance'. The figures for 1927-28 vividly illustrate the narrow basis of the net market production of grain which was of literally vital importance for the towns and therefore for the Soviet régime.[3]

TABLE 1

Grain balance (food and feed) for 1927-28

	million (metric) tons	per cent of gross production
Gross production	73.1	100.0
Stock changes	0.8	1.1
	———	———
Available for utilization	73.9	101.1
Utilization on farms:		
for seed	12.3	16.8
feed	23.3	31.9
other productive purposes	2.9	4.0
	———	———
total productive purposes	38.5	52.7
for food	27.3	37.4
Net marketable surplus	8.1	11.0
	———	———
	73.9	101.1

Source: Adapted from *Piatiletni Plan* (1929) II, 341.

Just over one-sixth of gross production consisted of seed corn and more than one-third was needed for animal feed and other purposes, e.g. for vodka distillation. Less than half the gross quantity harvested was available for feeding the peasants and for net sales to the non-agricultural sector. (Gross marketing from farms was substantially higher, because the peasants in some regions were not self-sufficient in grain production for their own needs.) The net marketable surplus of only 11 per cent of gross production had to feed the towns and the army, to supply fodder for horses and grain for industrial uses and even provided a minute surplus for exports.

The grain balance provides a simple model of the analysis of Soviet agriculture as a whole. Russian agricultural statistics start invariably with *gross production* obtained by adding up the value of all crops and animal products produced in a given period. If all current inputs, including depreciation on fixed assets, are deducted from gross production the remainder is called *gross income*. This consists of two elements: the remuneration of labour and the *net income* available for investment, reserve funds and tax payments.[4]

For an adequate analysis of the relations between an independent peasant agriculture and the rest of society such a scheme is inadequate and some intermediate steps may be inserted. In the first place, agricultural gross production includes a substantial proportion of products consumed in the process, such as feed and seed; if these are deducted from gross production, the balance is known as *final output*. In addition to agricultural raw materials, industrial purchases are needed to produce the final output and when these are deducted, the remainder is called *agricultural net production*.

Agricultural net production forms the *gross income* of the agricultural population. Part of it—and in peasant Russia this was a very small part—has to be set aside to cover the wear and tear of fixed resources, leaving the producers' *net income from agriculture* as the balance. In an expanding economy, not all of this is used for consumption, because the gradual expansion of the scale of farming operations needs more seeds for enlarging the sown area and more feed for more livestock. Nevertheless, by far the greater part of the agricultural net income is destined for personal consumption. In post-revolutionary Russia, the agricultural net income was distributed amongst almost 25m. peasants, the majority of whom were mainly subsistence farmers. A high proportion of the net income was, therefore, consumed in kind on the farms of origin and the balance was marketed, partly for sale to other peasants and partly to the rest of society.

The statistics of the time do not permit an accurate estimate of the importance of each type of transaction, but they supply some useful global data. In 1927-28, the value of agricultural gross production, including an increase of 500m. rubles in the value of livestock, was 14,500m. rubles and that of net production 8,900m. This leaves 5,600m. rubles or 40 per cent of gross production for the value of material inputs, both agricultural and industrial. We know that industrial goods for productive purposes were 22.2 per cent of the total rural demand for industrial goods.[5]

Net market sales of agricultural produce were 2,900m. rubles or almost one-third of net production and one-fifth of gross production. This estimate excludes the transactions between peasants and measures trade between them and the towns. Assuming that perhaps 500m. of this was sold in exchange for industrial inputs, the equivalent of 6,500m. rubles was, therefore, consumed in the village out of the total net agricultural product of 8,900m., either for private consumption or for the expansion of future output. Capital investment in agriculture, very largely by the peasants themselves, was very substantial: in 1927-28 it amounted to 3,100m. rubles, about 11 per cent, of the total agricultural funds of 28,700m. rubles and almost half the total capital investment in the country.[6]

Though the agricultural economy of NEP Russia was simple and backward, there was much more to the problem of stimulating the marketing of agricultural produce than providing enough salt and cotton goods in exchange for grain and butter. Both the economic functions of the actual transactions and the peasants' motives in marketing part of their produce were fairly complex. At least five separate elements in the market produce (m) can be identified:

m_1 consisted of the proportion of the gross product exchanged for current industrial inputs—nails, tools, some fertilizers, etc

m_2 was that part of the gross income exchanged for industrial goods needed to replace worn out fixed resources—ploughs and other large equipment, some building materials

m_3 was that part of the net income exchanged for industrial consumer goods, such as cotton textiles

45

m_4 consisted of another part of agricultural net income which was not consumed privately but exchanged for industrial commodities forming part of new agricultural investment in the form of new equipment and buildings

m_5 consisted of part of the net income sold in order to pay taxes without any direct economic equivalent in return.

Conditions of Static Equilibrium

The categories of Volume II of Marx's *Das Kapital* dealing with the simple and expanded reproduction of social capital have been adapted to Soviet industrial practice under the headings of industries producing means of production (Group A) and those producing consumption goods (Group B).[7] In these terms, $m_1 + m_2$ and, where applicable, m_4 are exchanged for products of Group A and m_3 for products of Group B.

The question of agricultural taxation in the conditions of NEP Russia raises some difficult issues. To the extent to which it was a contribution by the peasants towards the cost of Government services, it involved the sale of part of the agricultural net product (m_5) without direct economic return. However, taxation could also to some extent be regarded as a rent imposed by the Government as ultimate owner of the nationalized land on the income derived from favourable natural conditions or location. This had been expressly reserved by the Soviet power in the original land nationalization decree of 1917.[8] Technically this differed from a tax which involved the marketing of part of the peasant-farmer's output; it was a levy on the surplus income derived from the sale, at market prices, of produce with a particularly low cost of production or marketing. However, it may be doubted whether this distinction was practically relevant to the taxation policy of the Soviets and whether it would have been comprehensible or acceptable to the peasants, if it had been applied in practice.

In conditions of static equilibrium or 'simple reproduction', market sales by the peasants would be needed to replace current industrial inputs (m_1) and to make good wear and tear (m_2), to buy consumer goods (m_3) and to obtain money for tax payments (m_5). By definition, there would be no need to finance expansion and m_4 would, therefore, be zero, unless the peasants were eager to hoard money which was, on the whole, unlikely to be a significant factor. The only ways in which it was possible to stimulate an increase in marketable surplus without a preceding rise in production was by providing a greater supply of consumer goods or by increasing rural taxation. The former solution would involve a contraction in the consumption in kind of the peasantry compensated by an increase in the consumption of purchased industrial goods; with the gradual improvement in agricultural productivity and the falling proportion, or even the absolute size, of the agricultural population this would have been quite possible and compatible with a rising standard of living. Increased taxation, on the other hand, would have caused a relative, or even an absolute, fall in living standards.

This oversimplified scheme indicates the choices available in a situation where agricultural output stagnates but the need for marketed farm produce increases: the government can either induce the peasants to provide it through offering them more non-agricultural consumption goods or force them to consume less.

In the late 1920s, the elbow room for the extraction of more agricultural market produce on the basis of (more or less) static agricultural production had become very limited and the resources available for inducing the peasants by material incentives to make it voluntarily available were slim. The only hope of avoiding serious conflicts between the Soviet power and the peasants was a quick and substantial rise in output.

The Conditions of Agricultural Expansion
Agriculture may be required to contribute to economic development in a variety of ways, according to the Food and Agriculture Organization of the United Nations: by supplying food for the growing urban population and raw materials for the processing industries; by earning foreign exchange through exports; by providing a base for industrialization and by supplying capital and man-power for industry.[9]

The Soviet Government did not intend to base industrialization on agriculture, but with this exception the peasants were called on to contribute to Soviet economic development under all these headings. The conditions for obtaining such a contribution through rational measures were an increase in the volume and an improvement in the efficiency of agricultural production.

In practice, such results required more industrial supplies, both through current inputs such as fertilizers and through capital investment. In order to make the necessary intensification of effort acceptable to the producers, some increase in the supply of industrial consumer goods was also needed, though with the reduction in agricultural man-power through greater productivity individual incomes could be raised without a proportionate increase in the total supplies made available to the village.

The dilemma of the Soviet power was that it was in urgent need of the fruits of agricultural expansion but unable to supply the necessary means. The Communists knew that the primitive agricultural system which they had inherited—and which had been further weakened by the virtual destruction of large-scale estates—was more likely to act as a restraint on economic development than as a generator of economic growth. They had good reason to believe that the maximum benefit obtainable from such a system was that output might 'keep up with population growth, if the rate of urbanization is low. But agriculture is less likely to meet the needs of a growing population at the same time as urbanization is proceeding rapidly. The problem is compounded by increased *per caput* demand for food resulting from income growth and high income elasticities of demand.'[10]

Towards the end of the 1920s, the area under crops was virtually back to the pre-war level, and there were no reserves of fertile land which could be made productive without heavy capital investment. Nor did the country have the necessary infrastructure to make use of 'unconventional inputs' in the shape of 'research programmes, administrative structures for administering supply programmes for new forms of input and education in its various forms'.[11] Even if it is accepted that improved 'know-how' can quickly raise average production standards and the level of output, it depends on previous advances in the non-agricultural sectors of the economy. In countries struggling with the problem of how to make such advances, technological knowledge and administrative *expertise* for a concentrated attack on agricultural backwardness are in extremely short supply.

The Soviet planners had made absurdly over-optimistic assumptions about their opportunities for raising output without heavy investment, and particularly their yield forecasts were pathetically out of line with the facts. At first this may have been due to the rejection by the Government of more modest 'variants', but the same process continued long after the true effects of the mass collectivization policy had become only too apparent. The First Five Year Plan (1928-33) budgeted for a rise in average grain yields from 7.6 quintals per hectare to 9.5; in fact, the actual yield in 1932 was only 6.6 quintals and in 1933 6.9.[12] The official target of the first post-war Five Year Plan (1946-50) was a grain yield of 12 quintals, a figure well above any achieved before 1966 and maintained for the first time during 1966-68.

Painless expansion of agricultural output through the adoption of modern techniques available in advanced countries does not occur in the early stages of economic development when it would be most desirable. In the Soviet Union, the belief in this convenient *deus ex machina* provided a number of paper solutions to intractable problems but invariably led to grave disappointments.

In order to expand agricultural output quickly, new resources both from outside and from within agriculture are needed. Agriculture must supply labour, additional seed—which at the low grain yields usual in the Soviet Union at the time amounted to no less than one-sixth of gross production—and additional feed for the maintenance of larger herds. There were only two ways in which additional material resources could be made available: a cut in the consumption in kind by the peasants at a time when they had to redouble their efforts, or a cut in the marketable product.

The industrial inputs needed for additional production were partly current items, such as fertilizers, and partly more working and fixed capital. If they had to be exchanged for agricultural market produce, this involved either an increase in market production or a replacement of purchases of consumer goods (m_3) by producer goods (m_4). Alternatively, if the marketable product as a whole and its constituents could not be manipulated at will, the Government would have had to provide the additional resources without an immediate equivalent in the form of agricultural produce.

From whatever angle it is looked at, agricultural pump-priming implies a sacrifice of immediate personal consumption by the village, or by the towns, or by both. Either the marketable produce has to be increased at the expense of consumption in kind, or industrial producer goods have to be provided by the towns without a corresponding increase in market produce. Only after investment has led to greater final agricultural output will it be possible to do all things at once. The agricultural market produce will increase sufficiently to buy higher current industrial inputs (m_1), more producer goods for new investment (m_4) and more manufactured consumer goods (m_3), while the yield of agricultural taxation will also increase (m_5). At the same time, sufficient resources will be retained on farms to produce the larger final output, with something left over to improve the food supply of the peasants themselves.

The long-term strategy of the Soviet Government for the achievement of these aims was a combination of heavy capital investment of a kind quite new for Russia with elements of 'unconventional inputs' on a large scale. One of the basic tenets of the Marxist approach to agriculture, common to Lenin and Karl Kautsky, was the need for raising the productivity of agricultural labour through the replacement of small peasant farming by large, modern estates. In the primitive Russian village strip farming with traditional crop rotations was still the rule and the prospect for increasing output through reorganizing the cultivation system seemed excellent, provided the peasants could be induced to co-operate.

The encouragement of agricultural producer co-operation in general and 'collective' co-operative farming in particular, was thus seen as the key to modernizing Russian agriculture, but a more effective force than the slow influence of education and example was needed. This was believed to be the mechanization of the main cropping operations, which was regarded as an infallible way of convincing the peasants of the superior merits of the new system; thus the tractor became the symbol of this policy and the main agent of change.

Though generally ascribed to Lenin, the germ of this policy

goes, in fact back to Marx himself, who wrote as early as 1881 'that in the exploitation of the jointly-owned meadow lands the Russian peasants already practise the collective mode of production; that their familiarity with the *artel** would greatly facilitate the transition from agriculture by individual plot to collective agriculture; that the physical configuration of the Russian soil demands combined mechanical cultivation on a large scale . . .".[13]

Even after the experience of the last forty years it is impossible to quarrel seriously with this diagnosis—or with the conclusion that the encouragement of producer co-operation on the largest possible scale would have been essential to permit the Russian village to play a dynamic part in the economic development of the country. This was undoubtedly the solution which Lenin had in mind during the last few months of his active life, when he insisted that the palpable effects of mechanized agriculture would induce the peasants to join co-operative farms as a means of improving their living conditions quickly and substantially.

During the 1920s agricultural co-operation was a depressed industry. Co-operatives consisted mainly of poor peasants with little to contribute to the common fund except their uneconomic plots and their bare labour. If co-operation was to catch on as a voluntary movement, it needed a strong material lever in the form of tractors and combine harvesters which would make it possible to mechanize grain cultivation, the basic operation of Russian agriculture. Progress in agriculture thus seemed to depend on prior progress in industry which was to provide the plant and spare parts, the fuel and even the skilled man-power needed to operate the new equipment. The vicious circle thus looked as forbidding as ever.

However, the practical course of events may be expected to be less rigid than such a contrast suggests and the planners behind the first Five Year Plan were probably right to regard 'collectivization' through the formation of more producer co-operatives as an important method of raising yields and expanding marketable production. They therefore set themselves the maximum objective of raising the sown area of collective farms from 0.9

* Handicraft co-operative.

per cent of the total in 1927-28 to 14.3 per cent in 1932-33.[14] The transition to co-operation thus appeared to them a process covering one or two decades and they had no idea of the avalanche of crude violence and bureaucratic mismanagement which was to descend on the Russian village in 1929 and 1930.

The planners were similarly modest in their proposals to supplement partial collectivization with publicly-owned 'grain factories': their maximum aim was an increase in the sown area of State farms from 1.1 per cent of the total in 1927-28 to 3.5 per cent five years later.[14] In the event, the leadership again insisted on demanding the impossible and succeeded only in squandering large resources on the pursuit of a mirage.

In Professor Rostow's reading of economic history, 'agriculture must supply expanded foods, expanded markets and an expanded supply of loanable funds to the modern sector';[15] in Soviet Russia, the modern sector was expanded despite the drastic fall in food production and by starving the village of all but the most essential industrial products; loanable funds did not enter significantly into the economic nexus between town and country, but agriculture had to contribute towards industrialization its whole surplus over and above the bare necessities of life of the peasants—and sometimes even more.

2. AGRICULTURE IN AN INDUSTRIALIZED SOCIETY

When Stalin launched the first great industrialization drive in 1928-29, the Soviet Union was a mainly agricultural country with an important but limited and old-fashioned industrial sector. Economic advance was achieved through the ruthless concentration of investment efforts on heavy industry at the expense of the rest and this meant, above all, at the expense of agriculture and the peasants. When Stalin died a quarter of a century later, the Soviet Union had become a major industrial power. Its further development was no longer dependent on the extraction of the largest possible surplus from agriculture, but the growing imbalance between industry and agriculture threatened to impede the prosperity of the country and the process of economic growth as a whole.

Some of the functions of agriculture in relation to the rest of the economy are, of course, the same in an advanced and in a developing country, but there are significant differences of emphasis and balance. Particularly in a developed country where permanent reliance on food imports is ruled out, whether for political or for economic reasons, these functions may be summarized as follows:

(i) to serve as a source of man-power for the expanding sectors of the economy;

(ii) to feed the non-farm population;

(iii) to supply raw materials for certain processing industries, and

(iv) to provide, if possible, an exportable surplus to help to pay for imports.

The Supply of Rural Labour to Industry

During the 1930s industrialization was accompanied by a sharp decline in the total rural population (which may serve as a crude approximation for the agricultural population) from 124.7m. in 1929 to 114.5m. ten years later. The incorporation of new territories raised this figure to just over 130m. but the huge war-time losses in population fell heavily on the rural population and reduced it to only 109.1m. in 1950; since then it has been fluctuating around this level with only a marginally declining tendency.[16]

The number of people employed in public agriculture (State farms and collective farms, including fishing co-operatives and forestry) fell from 31.3m. in 1940 and 30.7m. in 1950 to 29m. in 1960 and 27.3m. in 1964[17] or by less than one half per cent per year in the 1950s and well over one per cent per year afterwards. These figures ignore, however, the individual peasants who were still appreciable in numbers before the completion of collectivization in 1949 and the family members of collective farmers and State farm staff employed on the subsidiary private plots. They, therefore, considerably understate the decline in the total agricultural labour force since the Second World War.

The proportion of the active population engaged in farming remains, however, very high for an advanced industrialized country. In 1964, collective farmers and their families employed in agriculture, whether public or private, constituted 22 per cent of the active population and the eight million workers and employees of State farms, together with the members of their families working on private subsidiary plots, raised this proportion to fully one-third of the total labour force[18]—only marginally less than the combined strength of industry and construction. This suggests that agriculture still represents a huge and barely tapped man-power reservoir for the expanding sectors of the economy, but it will depend on the rate of growth of productivity of agricultural labour how quickly, and to which extent, it will become readily available for redeployment elsewhere.

One cause of the still low productivity of labour in Russian agriculture is its acutely seasonal character, particularly in tillage. In 1959, the number of collective farmers and their families working in public agriculture varied from 18m. in January to 30.7m. in July;[19] the number of hands available for the harvest has to be supplemented by a temporary influx of workers from the towns but there was and is a great deal of rural unemployment throughout the autumn and the long Russian winter from November to April.

As the seasonal peak in labour needs is particularly steep in arable farming, its progressive mechanization and the growing importance of animal husbandry should gradually improve the seasonal utilization of labour in farming proper. At the same time, considerable attention seems to be given at present to providing more non-farming work for the rural population. The traditional crafts practised by the peasants during the winter were virtually destroyed during and after collectivization, but rather than falling back on the seasonal migration of labour (which also has a long tradition in the country), attempts are being made to bring industries to the village which can make use of the seasonally unemployed labour force.[20]

Feeding the Towns

Between 1929 and 1953 the urban population rose from barely 29m.—almost exactly the same figure as before the First World War—to 80m. people; twelve years later it reached 123m. and comprised more than 53 per cent of the total population. Agriculture still employed one-third of the population and every person employed in it had to feed, in addition to himself

and the non-working members of his family, at least two others
—a modest enough task according to the standards of other
advanced countries but a complete reversal in a single generation of the traditional balance of Russian society.

On the very crudest basis, the supply of marketable produce
for feeding the towns ought to have increased at the very least
in proportion to the increase in urban population, i.e. threefold
between 1928 and 1953 and over fourfold by 1964, in order to
supply the same average food basket per head of the population. Such a very rough estimate ignores, of course, a number
of important factors such as changes in the non-farming rural
population, foreign trade in foodstuffs, changes in consumption
habits, etc, but is nevertheless a useful first approximation.

TABLE 2
Market production of foodstuffs—1927-28–1964 (m. tons)

Item	1927-28	1940	1953	1964	1953 in % of 1927-8	1964 in % of 1927-8	1964 in % of 1953
Grain	8.1	38.3	35.8	74.1	442	915	207
Sugar-beet	9.8	17.4	22.9	76.1	234	780	332
Sunflower seed	2.3*	1.9	2.1	4.1	—	—	200
Potatoes	2.7	12.9	12.1	16.6	448	615	137
Vegetables	5.0	6.1	5.1	10.4	102	208	204
Meat and fat	1.4	2.6	3.2	5.8	229	414	181
Milk†	5.2	10.8	13.7	34.2	263	658	250
Eggs ('000m.)	3.9	4.7	5.8	11.3	148	290	195

* All oil seeds.   † Includes the milk equivalent of milk products.
Sources: 1927-28: *Piatiletni Plan* (Moscow 1929) II, 340. 1940 and 1953:
Narkhoz 1962, p. 233. 1964: Narkhoz 1964, p. 253.

By the time of Stalin's death, only potatoes and grain showed
a higher rate of increase in market production than the urban
population; supplies of all livestock products, vegetables, sugar
and probably vegetable oils had risen markedly less and the
supply per head of the urban population must have declined,
though not necessarily in exactly the mathematical proportion
between the rate of increase in supplies and that in the size of
the town population. Despite the large and broad-based expansion in market supplies during the following decade, gross
market supplies of vegetables and eggs rose considerably less

between 1928 and 1964 than the urban population, meat and fat just kept in step with its increase, while milk and milk products and the staple vegetable products had risen considerably faster.

In 1927-28 the supply of basic foodstuffs to the town population had been reasonable, though probably not particularly varied: 179 kg of cereals and cereal products per head per year, 49.1 kg of meat, 218 kg of milk and milk products, almost 91 eggs[21] and fairly plentiful supplies of vegetables were adequate for work and health even in rigorous climatic conditions. At the end of the Stalin era the composition of the diet had sharply deteriorated, with all animal products, vegetables and sugar in considerably shorter supply and a correspondingly heavier reliance on bread and potatoes than at the end of the NEP period.

During the Khrushchev era food supplies improved substantially, though at its end the position was still very uneven. Market supplies of grain, sunflower seed (the most important source of vegetable oil), vegetables, meat and eggs about doubled between 1953 and 1964, the expansion of milk and milk products and the raw material for sugar production was considerably greater and the only commodity where market supplies rose relatively little was potatoes. The improvement from year to year, though with a grave setback towards the end of the period, was plain for all to see, but the absolute level remained far from satisfactory. In 1958, bread and potatoes supplied 61.6 per cent of the total calorie intake per head of the population as a whole and only 21.9 per cent was derived from animal sources;[22] this suggests some improvement during the preceding thirty years, for in 1926 animal food provided only 13.8 per cent of the calorie intake of adult workers whose diet may well have been somewhat better than the average.[23] Nevertheless, the nutrition of the Russian people remains qualitatively well below that of other economically advanced countries.

A detailed statistical study of total food consumption between 1959 and 1963 indicates the extent of the shift both from basic foodstuffs to processed foods and from vegetable products to livestock products. During this period the consumption of foodstuffs derived directly from agriculture without processing by the food industry (i.e. mainly consumption by the farm population and food bought on the collective farm market) declined by 1 per cent (at constant prices), while purchases from the food industry rose by 30 per cent (and total food consumption by 20 per cent). All sub-groups of the food industry—bread and flour products, meat, fish and milk by about one-third, sugar by almost one-half—increased substantially, except flour, etc, and the fall in consumption in this case seems to represent mainly a shift to prepared flour products; nevertheless, in 1963 bread, flour and flour products still accounted for almost 18 per cent of the total personal expenditure on food.[24]

The pressure for more and better food, and particularly for the more expensive protective foods, is strongest during the early stages of the rise in living standards usually connected with economic development and the resulting change in the structure of the population. In the Soviet Union, the single-minded concentration on the development of heavy industry during the 1930s, the war and its aftermath of industrial reconstruction during the 1940s left this demand virtually completely unsatisfied during the Stalin era and its intensity afterwards was (and is) correspondingly greater. In its consumer aspects, including pre-eminently its feeding standards, the Soviet Union still remains a poor country compared with Western Europe, not to mention the United States, though it is well in advance of the underdeveloped nations east and south of its borders.

In most developed countries where living standards have increased substantially since the Second World War, the demand for foodstuffs is expanding only very slowly—and much more for processing and distribution services than for mere quantity—while technological advance tends to raise production levels higher and higher. The key issue in Western agricultural policy is an imbalance between excessive supply and sluggish demand, with periodical or chronic surpluses; in the Soviet Union it is, on the contrary, the lagging behind of supplies and the unsatisfied demand for the more valuable types of food which causes the greatest problems.

Agriculture and Consumer Goods Industries
The backwardness of the Soviet Union as a consumer goods

society is reflected in the very high share of food in consumer outlay and the correspondingly low proportion of industrial consumer goods. The official breakdown of all 'non-productive' expenditure, including personal consumption of material goods as well as that of public bodies and depreciation on fixed non-productive assets such as housing, provides a valuable statement of the overall position:

TABLE 3

The structure of consumer goods consumption 1959 and 1963

| | at current prices in % of total | | | | | | at constant prices % change 1959 to 1963 ('Total' only) |
| | 1959 | | | 1963 | | | |
	Total	Personal	Other	Total	Personal	Other	
Foodstuffs:							
(i) from farms	18.1	17.6	0.5	16.6	16.0	0.6	− 1
(ii) food industry	37.9	36.6	1.3	40.3	38.8	1.5	+30
Total food	56.0	54.2	1.8	56.9	54.8	2.1	+20
Other goods	38.9	33.5	5.4	37.0	31.0	6.0	+25
Depreciation	5.1	2.9	2.2	6.1	3.0	3.1	+48
Total	100.0	90.6	9.4	100.0	88.8	11.2	+23

Source: Narkhoz 1964, pp. 586ff and calculations therefrom.

These figures are not fully comparable with the data on final consumption in the national income statistics of Western countries, particularly in respect of services but they are, nevertheless, of considerable interest. In the present context the outstanding feature is the predominance of expenditure on food in personal expenditure on all goods. Though the physical increase in food consumption between 1959 and 1963 was somewhat lower than that of industrial consumer goods (and very much lower than the depreciation on consumer durables which in this connection include mainly housing), the share of food in total personal expenditure actually increased due to a rise in their relative, and in certain instances also in their absolute, prices.

Personal expenditure on industrial products in 1963

amounted to some 38,000m. rubles, of which products of the light industry (textiles and leather goods) absorbed no less than 22,700m. or almost 60 per cent. This branch shares with the food industry the distinction of having by far the highest share of raw material costs in its total costs of production—over 85 per cent in 1965.[25] Like the food industry it is mainly based on agricultural raw materials, though with the use of synthetic products and the gradual substitution of more thoroughly processed articles for simpler goods (e.g. ready-made clothing for cloth) the proportion of genuinely agricultural elements in the value of finished output is on the decline.

There has, of course, been a substantial rise in absolute terms in the output of industrial consumer goods since the Stalin era, when it was deplorably low. Supplies in 1964 were on average over three times the 1950 level and much higher for consumer durables which were in many cases virtually unobtainable at the earlier date. Nevertheless, these branches of industry have remained outstandingly weak for a country which registered some spectacular advances during this period in the most complex and expensive branches of modern technology. For the few products substantial enough to figure individually in official output statistics per head of the population, the level of supplies was as follows:

TABLE 4

Output of selected industrial consumer goods
(per head per year)

	Unit	1952	1958	1962	1964
Cloth (all types)	meters*	31	36	38	39
Hosiery	pairs	3.1	4.3	4.7	5.4
Knitted goods	pieces	1.6	2.4	2.9	3.4
Leather footwear	pairs	1.3	1.7	2.06	2.06
Clocks, watches	pieces	0.056	0.12	0.118	0.125

* Linear.

Sources: 1952-62: Narkhoz 1962, p. 127. 1964: Calculated from Narkhoz 1964, pp. 133f and population figures.

The rising purchasing power of the population creates a voracious demand for more and better consumer goods in all parts of the country. The satisfaction of this demand is impeded by a number of different factors. Fixed investment in consumer goods industries has been kept relatively low as a result of concentrating efforts on heavy industry and the highly capital-intensive modern defence industries. This has limited the productive capacity of the processing industries which could easily become a bottleneck preventing the full utilization of the results of a more successful agricultural policy than in the past. Another limiting factor has been insufficient concern for consumer demand in the choice of types and sizes of consumer goods and the subordination of customers' requirements to the convenience of the managers of industry, leading to excess production of unwanted goods and involuntary stock formation at a time of continuing overall stringency.

Nevertheless, the most prominent cause of insufficient supplies in the consumer goods industries has so far been the shortage of the basic materials supplied by agriculture. The only one which has not been a severe brake on production has been cotton, and industries processing livestock products (such as wool or hides and skins) are frequently hampered by the insufficient development of animal husbandry on which they rely for their essential raw materials.

The Changing Balance of Foreign Trade

Before the First World War, Russia exported very largely food and agricultural raw materials; imports consisted, in addition to tropical beverages and some luxury foods, mainly of certain raw materials and of manufactures. In 1913, grain exports exceeded 9m. tons, more than 10 per cent of the bumper crop of that year and about 40 per cent of the total quantity marketed.[26] This large surplus over domestic requirements was not simply due to a very favourable season; it also reflected the artificial limitation of domestic demand through the low living standards of the population, while an appreciable agricultural sector consisted of the large estates which produced largely for the world market.

Export availability was thus directly related to the dominant social and economic system and disappeared at first completely after the revolution of 1917. Grain exports remained low during the 1920s and in some years there were, in fact, substantial net imports. After the destruction of the social balance embodied in the New Economic Policy grain exports rose temporarily and the violent upheaval of these years was, indeed, partly caused by the struggle between the Government and the peasants about the disposal of the grain surplus.

In the last full year of the Stalin era, food and raw materials (including oil, ores and metals) still accounted for 40 per cent of all exports; this proportion was broadly maintained into the 1960s because a rise in the exports of non-agricultural raw materials balanced the steady decline in the relative importance of exports originating in agriculture.[27] In absolute terms, however, the quantities of grain, butter, sugar and cotton fibre sold abroad showed little tendency to fall and in some cases actually increased.

This was particularly true of the most important single agricultural commodity, grain. In 1952 grain exports amounted to 4½m. tons or almost exactly half the 1913 level; by 1958 they had risen marginally to 5.1m. tons and the excellent crop of that year raised the 1959 total to 7m. tons. Though the next three seasons were, on the whole, not very favourable to grain production, exports edged further upwards and in 1962 they reached their peak with over 7¾m. tons.[28] This rise was entirely due to growing exports of coarse grains, for wheat exports declined gradually from over 6m. tons in 1959 to about 4¾m. tons in 1962;[29] during the same period other grain exports thus trebled in quantity.

The crop failure of 1963 created a new situation: gross exports of grain fell to 6.3m. tons in that year and to only 3.5m. tons in 1964,[30] when exports of wheat were as low as 2m. tons.[29] At the same time massive imports of wheat began to arrive at Russian ports; they exceeded 3m. tons in 1963 and reached 7.3m. tons in 1964, plus the equivalent of a further 1.2m. tons of wheat in the form of flour. In 1965 and 1966 the Soviet Government bought a further 17 to 18m. tons of wheat from Canada, France, Australia and Argentina on contracts with delivery dates as far ahead as the summer of 1968.[31]

The only other foodstuffs exported by the Soviet Union in recent years in appreciable quantities were butter and sugar. Butter exports in 1962 were almost 70,000 tons, or only moderately less than the 78,000 tons exported in 1913.[28] They fell to only 25,000 tons in 1964 because of the effects of the 1963 disaster on milk production, but have since recovered sharply and have entered an already oversupplied world market in various forms and through various channels. Sugar exports reached the very high figure of almost 800,000 tons in 1962, increased still further in 1963 and fell back to less than 350,000 tons in 1964,[28, 30] the political character of this trade will be discussed later.

With agriculture taxed to the limit by the insatiable internal demand for food and raw materials, it might be expected that the Soviet Union is well on the way towards becoming permanently a massive food importer. The huge grain purchases since 1963 would fit into such a picture as an indication of the shape of things to come; in fact the situation is much more complex and less clear-cut.

There is no reason to believe that the Soviet Union will ever be able to combine progressive industrialization with agricultural expansion on American lines and to run a large export surplus on agricultural trade. In the years before 1964 the normal pattern of agricultural trade seems to have been a broad balance between exports and imports in value terms, perhaps with a slight tendency towards an export surplus. The large wheat and flour imports of 1964, amounting to almost 520m. rubles,[32] and the curtailment of grain exports produced a very large import surplus in trade in agricultural products which may be estimated at 850-900m. rubles or about one-eighth of the gross value of all exports in that year. A similar, though less extreme, imbalance may have continued since and the effects of the large forward contracts for wheat imports may take a year or two to work themselves out.

Nevertheless, such a state of affairs is more likely to prove the exception than the rule. On the political level—which is probably decisive—the consequences of permanent dependence on the capitalist West for a key item of the national diet would not simply be impalatable but intolerable for the Soviet régime. Economically, the difficulties of expanding industrial exports to the Western world sufficiently to pay for large-scale permanent grain imports would be formidable, and barter deals are much more easily arranged for expensive industrial plants with fat profit margins than for imports of primary commodities. Even if the substantial, though not accurately known, Soviet gold production were to suffice for the financing of such imports on a regular basis—which is by no means certain—the Soviet Government would probably not regard this as its best use, quite apart from the incidental effect that an expansion of the gold base for international liquidity would smooth the path of economic expansion for the Soviets' great rival.

On a different but not necessarily less significant level, food exports play a considerable part in the trade relations between the Soviet Union and the other members of the Communist bloc in Europe. Eastern Germany and Czechoslovakia are industrial countries which rely to some extent on imports of basic foodstuffs from the East in exchange for industrial goods, each buying on average about 1m. tons of wheat per year from Russia. Even the traditional food exporters amongst the Eastern European countries—Poland, Hungary, Rumania and Bulgaria—import in total about three-quarters of a million tons of wheat from the Soviet Union.[29] Another product where political elements enter even more prominently into Soviet food exports is sugar. The action of the United States in cutting off raw sugar imports from Cuba after the Castro revolution compelled the Soviet Union to become the major customer of its new ally, and as a result of the large raw sugar imports from Cuba in the early 1960s the Soviet Union became a large exporter of refined sugar.

Thus there are strong political and economic reasons for the assumption that the Soviet Union will regard self-sufficiency, with a modest margin for exports, as a major aim in its agricultural policy; this relates, in the first place, to the temperate staple products grain, meat and milk, but it is also clear that the Soviet Government wishes to produce the greatest possible quantity of fruit, grapes and tea in climatically suitable areas and to reduce the dependence on imports of such products as much as possible. In addition, some exports of basic foodstuffs to other countries of the Eastern bloc will remain necessary for economic and for political reasons for a balanced trade exchange between the Soviet Union and these countries, and in order to prevent their reliance on the capitalist West for vital supplies which might weaken the economic links between them and the Soviet Union.

As for trade with the rest of the world, the events following the crop failure of 1963 and the indifferent harvest of 1965 have made it clear that the Soviet Union is both able and willing to enter the world market on a very large scale in case of need. The Soviet Government has also made it clear that it has no objection to supply special needs, such as those of the extreme Far East, from overseas sources if this is the most economic course of action. Conversely, the world market will remain a convenient, if generally unrewarding, way of disposing of large seasonal surpluses which may well arise again. The commodities where this is most likely to be the case on a substantial scale are grain, oilseeds and butter(fat).

PART TWO

The Past

V

Agricultural Policy, 1917-1928

1. WAR COMMUNISM AND THE VILLAGE

The opening years of the Soviet régime witnessed the first great trial of strength between the Communists and the peasants. This was not the result of deliberate policy on the part of the Government, nor of deliberate hostility towards it on the part of the peasants. The Bolsheviks had come to power on the back of the great wave of peasant revulsion against the First World War, which culminated in a rural upheaval of the first order. This peasant war was directed against the landlords and against the minority of more ruthless and energetic *muzhiks* who had contracted out of the *mir*, the ancient and decaying village community, with the help of Stolypin's pre-war legislation. The peasants as a class had every reason to sympathize with a régime which took their part against the landlords in this struggle.

The Communists, on the other hand, overcame their theoretical preference for large-scale farming to the extent of endorsing in advance all the actions of the revolutionary peasantry. Less than two months before the October Revolution Lenin had written: 'The peasants want to retain their smallholdings, to keep them within certain norms, periodically to equalize them. . . . Let them! No intelligent socialist would quarrel with them on this point.'[1] Considering that his party had only the choice between being swept away by the sheer impetus of the peasant war and profiting from its consequences, the conclusion was plausible enough. Lenin's actions as head of the newly-formed Soviet Government were at first fully in line with his earlier pronouncement. The Land Law passed by the new government immediately on grasping power, was, in Lenin's own words, 'entirely copied from the S.R.* platform'.[2]

* Social Revolutionaries.

However, within a few months the vital necessities of the régime in its struggle for survival against foreign and domestic enemies involved it in a grave conflict with the inherent individualistic tendencies of the peasant economy which foreshadowed in important respects the collectivization crisis of 1929-32

The economic imperative of War Communism was plain and inescapable. The Soviet régime was surrounded by a hostile world which favoured and supported the counter-revolutionary troops led by a succession of high-ranking Tsarist officers; in fact, the Western powers actively intervened in the uneven struggle, whenever this was technically and militarily possible. To avoid defeat at the hands of their infinitely better equipped and supplied opponents, the Soviets had to throw all their resources into the struggle and to mobilize all their allies and sympathizers. These were to be found, above all, amongst the industrial workers of the starving towns where production was inexorably grinding to a halt. Very little of the remaining tiny output of manufactured goods was available for sale to or barter with the peasants in exchange for the food on which the red soldiers and the town population depended for survival from day to day. The needs were even more desperate, because for a long time the traditional granaries of the country—the Ukraine, southern Russia and western Siberia—were denied to the Soviet power.

The Communists were thus thrown back on a policy of systematic requisitioning, for payment in worthless paper money at absurdly low prices did not transform one-sided spoliation into a genuine exchange of equivalents. In applying this policy they attempted, generally with little success, to make use of the internal social divisions in the Russian village into poor peasants, middle peasants and *kulaks*. Immediately after the radically egalitarian revolution of 1917 these divisions were much less significant than before, and the official organization of 'committees of the village poor' (complete with a national congress of such committees held in Petrograd in 1919) did little to reduce the solidarity of the peasants in face of outside pressure.

From the peasants' point of view, compulsory requisitioning was, of course, rank robbery, and the official benevolence of the

Soviets towards the middle peasants was probably more than balanced by Lenin's emphatic declaration that 'free trading in grain means freedom for the capitalist' and would not return.[3] The only reason why the peasants put up with this system as long as they did was their instinctive grasp of the fact that the White Guards under their ex-Tsarist commanders were their worst enemies. The final defeat of the counter-revolutionary armies was, therefore, followed very quickly by peasant risings against the practices of War Communism.

From an abstract point of view, the Communists were undoubtedly fighting the peasants' battles as well as their own in waging war against the armies of Kolchak and Yudenitch. However, this broad similarity of aims could not, and did not, influence the reaction of the individual peasant confronted with repeated, and increasingly heavy, demands from his overbearing allies and protectors. If it was a question of seeing his farm animals requisitioned in order to feed the towns or slaughtering them himself, the peasant preferred to kill them for himself and to feed himself and his family; if the result of growing grain and storing the surplus was to have his store raided and his grain confiscated, he would reduce his sown area, partly from necessity and partly from choice, to the bare minimum needed for his own requirements.

The result was a vicious circle of steadily decreasing agricultural output. In 1917, agricultural gross production was about 12 per cent below the level of 1913. As the base year had been unusually favourable for crops, even the decline of 19 per cent in their volume was, though serious, not intolerable, particularly because there were no enforced exports. Livestock products were estimated, perhaps over-optimistically, to have remained at the pre-war level. During the following three years, the losses caused by the peasant risings and the effects of Civil War and Communist economic policy depressed agricultural gross production by a full quarter to only two-thirds of the 1913 level, and this loss was spread almost evenly between crops (down to 64 per cent) and livestock products (72 per cent of 1913) (Appendix, Table I). Gross grain production in 1920 was only 54 per cent of the average for the years 1909-13 and for wheat and rye, the two most important 'bread grains', the decline was even greater, while the production of cotton and sugar beet fell to negligible levels.[4]

Such figures, even if accurate, can give only the very palest idea of the contraction in farm output, and in the chaotic conditions of the time the collection of reliable statistics was, of course, out of the question. The fall in livestock numbers is believed to have been comparatively modest: between July 1916 and 1920 the number of horses, cattle and pigs declined by about 14 to 16 per cent and that of sheep and goats by 6 per cent.[5] These losses were substantial, and in the nature of things not quickly replaceable, but by no means catastrophic.

While agricultural production fell to danger levels, the gradual increase in the adminstrative efficiency of the new bureaucracy and the recovery of enemy-held territories were reflected in a steady increase in the volume of grain collected for the army and the towns. From a low point of 108m. pood (under 1.8m. tons) in 1918 it rose to $212\frac{1}{2}$m. in 1919 and 367m. in the following year.[6] 'But this great organizational success carried in itself the seeds of its own destruction. The real surplus of Russian agriculture had completely disappeared a long time ago: the quantities of foodstuffs collected by the Soviets increased at the expense of the nutrition of the peasants and the seed grain for the coming year . . . The time was swiftly approaching when the towns could be fed only at the expense of next year's crops. The existing stores had been consumed without being replaced by new reserves, peasants and requisitioning troops made big inroads into the seed grain for the coming sowing campaign—and famine impended in towns and villages alike. This was the state of things during the winter of 1920-21.'[7]

The only possible justification of the system of War Communism was the over-riding necessity of a military emergency which at that time no longer existed. The system had thus become unnecessary as well as self-defeating, and a way out of the impasse was found in the New Economic Policy. Considering the momentous nature of this step, which involved a radical change in the direction and methods of Soviet economic policy, it was taken surprisingly smoothly. The factors which produced this result were Lenin's unchallengeable authority and, reinforcing it, the severe jolt administered to the self-confidence of the

Communists by the Kronstadt rising, a rebellion of the revolutionary sailors—who had been their staunchest supporters—against the excesses of the bureaucratic dictatorship established as a result of the Revolution.

The basic pattern of the relations between the Soviet power and the peasants during this period was fated to recur more than once. It was a fundamental divergence between the immediate economic interests of the two groups, coupled with a real difficulty which compelled the Government, in the interests of self-preservation, to use extraordinary measures for its solution. These measures acquired a dangerous and excessive momentum of their own, until the whole of Soviet society was in the throes of a first-rate crisis which affected in the first place the peasants but ultimately the whole economic and political system.

2. THE NEP AND PEASANT INTERESTS

Discussions on the need for the abandonment of War Communism went on in the inner circles of the régime for the best part of a year, while the huge governmental machine of compulsion was grinding on until it had to be thrown violently into reverse. This was done, rather unceremoniously, when the 10th Congress of the Communist Party decided in March 1921 to end the requisitioning of farm produce and to replace it by a progressive tax in kind. Though on the surface a modest enough change, this measure was of cardinal importance as a renunciation of the previous policy of appropriating the whole surplus product of agriculture over and above the subsistence needs of the peasants: 'All stocks of food, raw materials and fodder remaining to the peasants after the discharge of their tax obligation are completely at their disposal and may be applied to the improvement and strengthening of their economy, the improvement of their personal consumption and to the exchange for products of industry, handicraft or agriculture.'[8]

The tax in kind was replaced in May 1923 by a mixed tax partly payable in cash and partly in produce and a year later by an ordinary monetary tax. It was at first levied over a wide range of commodities, including grain and potatoes, animal products of all kinds and even hay and straw. The rates were generally 50 to 60 per cent of the proceeds of the compulsory deliveries of 1920-21, but only at about a quarter of that level for meat and hides.[9] Thus it did not produce nearly enough food for the needs of the towns. In terms of agricultural gross production—which is, however, an unsound basis of comparison—it amounted to 2.8 per cent in 1923-24, 3 per cent in 1924-25 and only 1.8 per cent in 1925-26 when the rates were reduced as part of the conciliatory policy under the slogan of 'the face towards the village'.[10]

In the years immediately after the introduction of the NEP, the initiative in the process of getting the economy moving again was reserved for agriculture. In the first place, Russia was even more overwhelmingly agricultural under the Soviets than it had been before the war. Between 1917 and 1920 many starving town-dwellers had gone back to the countryside, where most of them still had friends and relatives, while many others perished from malnutrition and disease or were killed on service in the Red Army. Within these three years the urban population declined by almost five million, or nearly one-fifth.[11]

In the second place, the decline in the output of industry during the period of War Communism had been much steeper than that of agriculture; in 1921 industrial gross production was less than one-third of the 1913 level, while even in that year of famine agricultural gross production was about three-fifths of the pre-war figure.[12] The mercifully favourable season of 1922 furnished at least some spare resources for economic 'pump-priming', provided the Government could coax them out of the hands of the peasants to whom they accrued in the first place. The NEP thus proved very quickly its potential for producing the golden eggs which War Communism had tried in vain to obtain by killing the goose.

With the tax in kind providing not much more than half the meagre proceeds of compulsory deliveries under War Communism, the towns had to acquire food and agricultural raw materials from the farms in exchange for industrial goods. This meant greater claims on the consumer goods industries which were enabled by the growing supplies of raw materials to expand their output quite considerably. Between 1922-23 and 1923-24, industries based on agricultural raw materials doubled their output, while production in the rest of industry rose only by one-quarter. In 1923-24 the industries processing agricultural raw materials accounted for well over one-half of total industrial production—in the light of later developments a truly staggering proportion.[13]

The emphasis on market demand and market requirements involved a fairly thorough-going reshaping of the system of industrial organization as a whole. It also made it essential to establish a stable currency in place of the Soviet ruble which was in a state of miliary inflation. This was achieved through a currency reform based on the '*chervonets*'; the economic policy needed to keep the new currency in reasonable, though far from perfect, balance involved a system of financial discipline whose social cost was high, even though the orthodoxy of the Ministry of Finance was in practice tempered by the needs of industrial reconstruction.

The most important structural change in the Soviet economy was, however, not the deliberate result of government policy but occurred spontaneously. The retreat from the extremes of War Communism created the 'Nepman', a peculiar type of private business man, barely tolerated by the authorities but for the time being an indispensable agent for organizing a substantial part of the exchange between town and country. He was aware of the precarious and transitory character of his position and, therefore, concentrated mainly on trade where he could obtain a quick and lucrative turnover of his capital which was less at the mercy of the Soviet Government than fixed investment would have been. Running great risks he aimed at high and fast profits, and the absence of an efficient system of exchange between the nationalized industries of the towns and the village gave him the chance of earning them.

The huge number of small independent peasant holdings, operating generally on the verge of subsistence and disposing only of small and fluctuating surpluses for the market, needed the intervention of flexible middlemen, as the system of so-called consumer co-operatives was too clumsy and inefficient to organize the collection of agricultural produce in exchange for manufactures. As sellers, particularly as sellers of grain, the small peasants usually either sold to the local *kulak* or were at least strongly influenced by him in the timing of their sales. Wherever possible, they also sold to the private trader who offered them a better assortment of industrial goods in exchange than the public trading system. In the first full year of the NEP, more than half the marketable surplus of the village took the form of the tax in kind and only 44 per cent was sold on the market; this proportion was well over half for potatoes and oil seeds, eggs and poultry, but much lower for grain.[14] With the replacement of the tax in kind by a money tax, the whole marketable surplus of agriculture became the subject of a free sale, with part of the realization price used for paying the tax and the balance available for exchange against manufactures.

The closer the approach to a market economy, the greater the chances of friction between the three agents meeting in the market place: nationalized industry, the peasants and the Nepmen. Until after the relatively ample harvest of 1922, the purchasing power of food in terms of industrial goods was very high; the 'terms of trade' were in favour of agriculture. This was not surprising, because in the first year of the NEP (1921-22) food was desperately short and correspondingly dear. In addition, the strikingly inept methods of the Communists in adapting their industrial system to a market economy produced a huge bargain sale of manufactures at give-away prices, because public undertakings were not supplied with adequate working capital and had to sell off stocks in order to get enough liquid funds. The favourable terms of trade were of little use to the peasants who had nothing to sell but they permitted the Nepmen to buy up manufactures at low prices; their chance of making quick profits came in the autumn of 1922, when the peasants disposed of relatively large marketable surpluses and were eager buyers of manufactures for the first time since the October Revolution. This relative plenty of food and raw materials demonstrated the insufficiency of manufactured goods in the shops and created the basis for the 'scissors' crisis' of 1923.

The main symptom of this crisis was a growing disproportion between the prices of agricultural and industrial products, both at wholesale and at retail level.

In the autumn of 1922 and during the first half of 1923, the indices of prices of industrial goods at the retail level were rela-

TABLE 5

Ratio of industrial to agricultural prices
(1913=100)

		Wholesale	Retail
1922	October	131	161
1923	January	156	184
	April	190	221
	July	202	211
	August	241	187
	September	294	280
	October	310	297

Source: *Sovietskoye Narodnoye Khoziaistvo 1921-25*, p. 413.

tively higher, and often substantially higher, than at the whole-sale level: the private traders managed to keep well ahead of the increased prices charged by nationalized industry. From July 1923 onwards, the opposite began to happen: the State-owned trusts and syndicates were instructed to go out for the highest prices the traffic would bear, and during the autumn of 1923 the relative wholesale prices of manufactures actually exceeded the retail prices.

However sensible this may have appeared to the Supreme Economic Council, this policy—coupled with the effects of a relatively poor harvest—soon demonstrated the operation of the law of dangerous over-success in the relations between the Soviet power and the peasants. The purpose of the official policy was to extract the greatest possible volume of agricultural produce for the same (or a smaller) volume of manufactures; its real effect was a painful check to the developing exchange between town and country and a cruel setback to the New Economic Policy as a whole. Turnover in manufactures slumped and stocks accumulated, while the volume of some important agricultural commodities acquired through the public trading system actually declined below the level of the previous year.

This first trial of strength in terms of market strategy between the main social forces of NEP society contained in a nutshell some of the most critical and controversial issues affecting the further development of the Soviet economy, and of Soviet society as a whole.

TABLE 6

Purchases of agricultural produce through public trade

Product	unit	1922-23	1923-24	% change
Grain	000 t	6,919.1	6,527.2	− 6
Oil seeds	000 t	354.2	518.7	+ 46
Raw cotton	000 t	22.7	136.5	+501
Meat, cattle*	000 t	414.3	329.1†	− 21
Eggs	waggons	1,596	1,071	− 33
Wool	000 t	5.2	14.0	+169
Hides, large	000	4,042	3,975	− 2
small	000	4,740	5,555	+ 16

* Live weight. † 1924-25; no data available for 1923-24.
Source: *Sovietskoye Narodnoye Khoziaistvo 1921-25*, p. 448.

In the short run the crisis was overcome by a reduction of 25 per cent in the prices of industrial goods in 1923-24,[15] coupled with increases in the acquisition prices of grain and agricultural raw materials—a painful acknowledgment by the Government of the current balance of power within the economy. In the long run, the conclusion drawn from this economic defeat of the authorities by the 13th Conference of the Communist Party in January 1924 was probably more important. It formulated as one of the urgent tasks the elimination of the private trader from his position as intermediary between public-owned industry and the peasants. Accordingly the proportion of grain procured by private traders fell from 7.7 per cent in 1923 to only 2.2 per cent in the following year, and by the end of the post-war recovery period the Nepman was virtually driven from this important field.[16] This development was an early indication of the approaching end of the compatibility between the NEP as a whole and the growth tendencies of the Soviet system.

The next few years were, however, to demonstrate first the usefulness of the NEP framework for the recovery of production in the economy as a whole, agriculture as well as industry, to the pre-war level and somewhat higher.

3. THE RECOVERY OF OUTPUT AND ITS LIMITS

The vigorous growth in agricultural production during the early years of the NEP made it permissible to regard such early signs of friction between the Government and the peasants as little more than healthy growing pains.

The new policy could not prevent the outbreak of a grave famine in 1921-22, following a drought in the summer of 1921 on top of three years of War Communism, but the primitive subsistence agriculture of post-revolutionary Russia recovered remarkably quickly from the crisis. However unfavourable the long-term consequences of its backwardness, at a time of material poverty and social disintegration the far-reaching independence of the peasants of outside resources made it comparatively easy for them to re-establish their traditional routine. As soon as they could raise the necessary seed grain and muster a modicum of animal draft power, they got the temporarily abandoned land back under the plough and brought the production of crops up to the traditional level.

This process went on despite a desperately low supply of new tools and machinery. In 1913, the agricultural system as a whole bought 109m. rubles worth of these items, almost half of which was imported from abroad. This pre-war level was not exceeded until 1925-26; total supplies during the six years from 1918 to 1924-25 amounted to only 120m. pre-war rubles or ten per cent more than in the single year 1913. With an agricultural economy consisting of almost 25m. holdings, the number of ploughs produced during the first four years of the NEP was 674,000. These were supplemented by 191,000 harrows, 47,000 winnowing machines, 73,000 reapers and mowers and 71,000 threshing machines—altogether less than one major implement of any kind for one peasant household in twenty. Only from 1925 onwards did the output of agricultural tools and machinery make rapid progress: almost as many ploughs and harrows were turned out in a single year as in the preceding four years together.[17]

However, these improved supplies made little further impact on agricultural output which by then had regained its pre-war level and tended to stick there, particularly in the production of the main crops. (Appendix, Tables I and II.) With the exception of potatoes and oil-seeds, both of which had more than doubled, production clearly hovered around, and in the crucially important case of grain slightly below, the pre-war level, with the production curve flattening out completely.

One of the causes of this disappointing performance was the lack of any sustained improvement in yields. Apart from potatoes, yields were, indeed, generally below the pre-war level, though there were large fluctuations from year to year:

TABLE 7
Major crop yields 1909-13–1928
(in quintals per hectare)

	Grain	Sugar-beet	Potatoes	Sunflower seed	Cotton (irrig.)	Flax
1909-13	6.9	150	78	—	13.0	2.8
1913	8.2	168	76	7.6	10.8	3.2
1925	8.3	165	76.5	7.2	9.5	2.8
1926	8.2	118	82.5	6.0	8.5	2.4
1927	7.5	151	75.1	7.6	9.6	2.4
1928	7.9	132	81.8	5.4	8.5	2.4

Sources: 1909-13 and 1913: Selkhoz, pp. 196ff. 1925-28: Jasny, *Socialized Agriculture*, p. 791.

Comparisons with pre-revolutionary conditions are valid only in broad terms; for grain, 1913 was an exceptionally favourable year and the average of 1909-13 is, therefore, in principle preferable, though different figures are quoted in the literature.[18]

On the whole, the success achieved in crop production during the 1920s consisted in the restoration of pre-war output, perhaps with marginally lower yields, on a somewhat larger cultivated area, due to a considerable extent to a substantial increase in the acreage of the so-called 'technical crops' (cotton, sugar-beet, oil-seeds, flax, tobacco, etc.). By 1926 the index of crop production was 14 per cent up on 1913, but the situation was patchy and two years later it had crept up only to 117 (1913 = 100). (Appendix, Table I.)

On this almost stagnant base, animal production made comparatively rapid progress. It exceeded the 1913 level by 27 per cent in 1926 and by 37 per cent in 1928—the highest level it was to attain for the next quarter of a century. This more dynamic development was due to a steady increase of all types of productive livestock and to a structural shift compared with the pre-revolutionary situation from horses to cattle, pigs and sheep. (Appendix, Table IV.) The reasonably fast and quite steady growth in livestock numbers was the more remarkable, because differences in crops tend to be reflected, with some time lag, in the state and size of herds and flocks which depend largely on the feed available from the previous crop, and weather conditions cause wide annual fluctuations in crop levels. Partial droughts and crop failures occurred not infrequently during the NEP era but the national herd increased with hardly a setback. The war-time count of horses was never equalled, but cattle, pigs, sheep and goats all went up appreciably; total numbers exceeded the 1916 level as early as 1925, less than four years after a serious famine.

Although the agricultural production record of the period was thus on the whole tolerably good, the most critical, and in the long run the most ominous, fact was the stagnation of grain production. Ignoring the exceptionally good 1913 harvest, grain production during the later 1920s may have exceeded the average pre-war level by about 10 per cent but it showed no discernible upward trend towards improvements in yield or expansion in the sown area. It was, in fact, at the 'bread front' that the social and economic policy of the NEP period collapsed in 1928-29.

4. THE STRUCTURE OF THE NEP VILLAGE AND THE MARKET

In 1928, the total sown area amounted to about 113m. hectares, of which individual peasants were in possession of 108¾m.[19] As the number of holdings was estimated at slightly less than 25m., the crude average sown area per holding was therefore about 4.3 hectares or 11 acres, with an annual gross production of grain of less than 3 tons, potatoes 2 tons and a few hundredweights of cash crops. The average peasant householder had fewer than 1½ horses, three head of cattle (including about a single cow), 1 pig and 5 sheep and goats. The average unit of production at the end of the NEP period was, therefore pathetically small.

Such averages are of very little value for purposes of scientific analysis and they were also of much less immediate political interest than the range between different groups and the relative weight of small and larger units as suppliers of agricultural produce for feeding the urban population and keeping the processing industries busy. In the longer run, however, the diminutive scale of the average productive unit was a fact of the utmost importance: it implied that any significant advance in productivity required a structural revolution through the amalgamation of the excessive number of dwarf holdings into farms large enough to permit the application of modern technological methods.

The results of a large-scale sample investigation of some 600,000 peasant households in 1927 provide an extremely detailed picture of the prevailing agricultural structure. Its main drawback is, indeed, the excessive subdivision of the sample into seven main groups; as the average differences between some of them are sometimes very small, this tends to blur the findings to a certain extent.

TABLE 8

The structure of peasant households in 1927
(per household)

	Total or Average	Sub-groups						
		1	2	3	4	5	6	7
Value of means of production—rubles	516.3	none	up to 100	100-200	200-400	400-800	800-1,600	over 1,600
% share of total	100.0	3.0	10.7	12.3	26.3	30.9	13.6	3.2
Population per farm:								
average total	5.1	2.6	3.6	4.1	4.9	5.7	6.5	7.3
average men workers	1.1	0.6	0.8	0.9	1.1	1.3	1.5	1.7
% without men workers	13.6	42.6	30.9	21.0	12.6	7.4	4.7	4.4
Land tenure: (ha)								
allotted arable	6.1	2.5	3.7	4.5	5.6	6.7	8.8	11.0
land used—arable	4.0	0.8	1.5	2.2	3.3	4.7	6.9	9.4
meadows	1.2	0.3	0.5	0.8	1.1	1.4	1.6	1.7
total	5.2	1.1	2.0	3.0	4.4	6.1	8.5	11.1
% letting land	13.0	34.7	32.7	20.9	11.3	6.3	5.1	7.0
% leasing land	17.3	3.2	6.2	7.9	12.9	21.4	32.0	39.2
Livestock (units)								
Work stock	1.0	—	0.1	0.4	0.9	1.2	1.7	2.3
% without work stock	31.3	100	95.2	61.6	24.8	9.4	5.2	10.0
Cows	1.1	—	0.3	0.7	1.0	1.3	1.7	2.5
% without cows	23.8	100	70.8	33.2	19.5	9.4	5.9	7.8

Source: Jasny, *Socialized Agriculture*, pp. 780ff from Statistical Handbook of the USSR (1928).

Perhaps the most striking feature of the situation was the narrow range of the social differentiation of the Soviet village ten years after the Revolution. This was, indeed, very largely the result of the Revolution itself which had produced a very egalitarian land distribution. Not only are the differences between the various group averages small in absolute terms but they are associated with substantial differences in the average size of households. The differences per head were, therefore, even smaller; if the comparison is based on the number of male workers—which in a traditional society such as that of the Russian village of the 1920s probably produces the most significant results—the distribution appears to be even more egalitarian.

The details of land usage underline the extremely small scale of operations of the great mass of the peasantry. Even the largest of the three bottom groups had an actual land usage with an average of less than $7\frac{1}{2}$ acres per household, the two middle groups averaged less than 15 acres and the top group averaged less than 30 acres, including the land hired mainly from the poorer households.*

In population terms, the three poorest groups of households accounted for 26 per cent of the holdings but may have contained less than 20 per cent of the population, while the two top groups had under 17 per cent of the households and rather more than 20 per cent of the population. In terms of land usage, the three bottom groups cultivated less than one-eighth of the arable land, the two medium groups had about the same proportion of households as of land (slightly more than half) and the two top groups disposed of almost one-third of the cultivated land.

These differences suggest that the radically levelling forces at work during the revolutionary era were on the retreat but that the differentiation had not progressed very far. This may be illustrated by comparing the results of this rather sophisticated sample with the crude estimates of the pre-revolutionary period. These suggest that in 1913 the poor peasants were 65 per cent of all households, the middle peasants 20 per cent and that the

* This division corresponds to the contemporary distinction between poor, middle and well-to-do peasants.[20]

kulaks, with 15 per cent in numbers, accounted for 37 per cent of the land held by the peasants.[21] The definition of the *kulak* class was notoriously influenced by political considerations and the dividing lines were very rough-and-ready, but the reversal of the relative proportions of 'poor' and 'middle' peasants is a neat summing up of the effects of the Revolution on the Russian village.

One important area in which the incipient social redifferentiation of the peasantry was very noticeable was the way in which different groups of holdings either had to find outside employment for their surplus labour or absorbed the unemployed labour resources of others.

TABLE 9

Income structure and labour utilization of peasant households
(in % of households)

	Average	Sub-group						
		1	2	3	4	5	6	7
Source of income:								
wholly non-rural	2.8	24.6	10.9	3.6	1.2	0.4	0.2	0.7
partly non-rural	5.4	10.7	10.7	7.9	5.0	3.5	3.5	3.9
Hire and sale of labour:								
hiring labour	19.8	4.7	8.3	11.1	15.6	23.0	34.0	51.1
selling labour	35.4	58.0	58.4	48.2	36.6	28.0	20.4	11.5
exchanging labour	6.4	6.8	13.8	11.2	6.4	3.8	2.5	1.5

Source: As Table 8.

The two lowest groups thus contained a substantial proportion of households which relied on non-agricultural, and even on non-rural, sources for their total livelihood or part of it and more than half of the households in the three lowest groups depended to some extent on the sale or exchange of labour by their members. The two middle groups had very little non-rural income and in them the hiring of labour assumed some importance, though it was less frequent than the practice of going out to work for others. In the two top groups, and particularly the highest one, reliance on hired labour was increasingly common, though a proportion of such households still supplemented their income by working for others.

The structural and social differences in the NEP village of 1927, though by no means negligible, were thus far from spectacular. Fragmentary data for earlier years suggest that the larger peasants had made some headway at the expense of the poorest (but not of the middle) peasants even by 1925,[22] but in a sub-continent with huge regional and physical differences even the most thorough-going local equalization of land, such as took place in the wake of the 1917 Revolution, was bound to leave some differences between areas in the average size of land and livestock holdings. Some change towards social differentiation had undoubtedly taken place in the decade following the Revolution, and even without statistical significance tests the close association between average capital and land holding and such features as family size, proportion of holdings without male workers, outside sources of income and the hiring and selling of labour is at the very least highly suggestive.

The official interpretation of this evidence is a rejection of the Trotskyist emphasis on the growing *kulak* danger in favour of the theory that the NEP was a time of consolidation for the 'middle peasants'. It is probably more correct to say that the Trotskyists correctly diagnosed a trend but greatly exaggerated the extent to which it had already become a reality. However, if the growing difficulties of the régime in its relations with the peasants could not simply be attributed to the machinations of the 'class enemy', they were even more fundamental than the protagonists in the debate appreciated, for they expressed a basic incompatibility of interests. The essence of the difficulty was that for the peasants the Revolution was accomplished as soon as they had redistributed the land and shaken off the burdens which they had been compelled to bear under Tsarism; for the Communists, on the other hand, the October Revolution was the necessary condition for carrying out the economic revolution which was needed in order to transform backward peasant Russia into a modern industrial nation, while permitting them to remain in power as the effective government of the country.

The peasants wanted to be left in peace, to arrange their own affairs as they pleased and to be able to buy all the manufactures they wanted in exchange for the surplus of their production over their subsistence needs. There were obvious limits to the expansion of agricultural output in the primitive Russian village of the 1920s, but it is very doubtful whether these limits were approached in practice. This restraint on production was not caused by the peasants' preference for leisure compared with goods and even less by the saturation of their demand for industrial products in exchange for their own produce. The main reason for the brake on agricultural output was the 'goods' famine' which prevented the peasants from obtaining adequate quantities of manufactures at reasonable prices and thereby acted as a powerful and increasingly effective disincentive to larger market production.

From the Government's point of view this problem appeared as the insufficiency of agricultural market production for the needs of the towns, whose population had more than recovered from the losses incurred during the Civil War years and was growing by almost one million people per year. A much-quoted but misleading comparison between 1913 and 1927 indicated that, while gross production was almost back to the pre-war level—according to contemporary calculations it amounted to 4,749m. poods or about 78m. tons in 1926-27 as against 5,000m. poods or almost 82m. tons in 1913—the quantity of grain entering non-village trade had declined to less than half (630m. poods or 10.3m. tons as against 21.3m. tons).[23] This difference was partly the result of statistical errors and to the extent that there was, in fact, a genuine reduction in market production, it was largely due to the disappearance of the large estates; these had been strongly market-orientated and their abolition left the towns dependent on the fluctuating surpluses of very large numbers of small producers over their own requirements. Although the effect of the fall in market production of grains on the supply of the towns was partly compensated by the virtual cessation of grain exports, its potential consequences for the economy as a whole were extremely serious.

The fall in the quantity of animal products reaching the towns was also severe. Between 1913-14 and 1925-26 meat production for sale fell by 16.4 per cent, while milk supplies in the following year were 22.4 per cent, and those of animal fats 15.5 per cent, below the pre-war level.[24] Such reductions were far from negligible, particularly because for these commodities the cushion of pre-war exports was less substantial or completely

84

missing and the decline was felt mainly by the urban consumers.

It is the great paradox of this period that in the 'first workers' state in history' the market became the great battlefield for the conflicting interests of the Soviet power and the mass of the rural population and that the price and supply movements on the agricultural produce markets became the clearest barometer of social and economic developments.

The prices obtained by the peasants during the later NEP years for different commodities showed a wide divergence between crops and livestock products, with the all-important grain prices changing in a striking fashion from year to year.

TABLE 10

Planned procurement prices 1924-25–1927-28

(producer prices 1909-13=100)

	1924-25	1925-26	1926-27	1927-28
Grains	137.0	137.5	108.0	115.0
Rye	128.6	135.5	100.2	106.6
Wheat	142.9	141.3	113.6	117.2
Oil-seeds	89.3	94.5	100.4	116.9
Other technical crops	143.0	138.3	135.3	139.5
Animal foods	123.8	159.8	171.2	173.0
Raw materials of animal origin	143.2	159.7	171.9	177.7
All farm products	132.8	143.7	136.9	142.8

Source: Quoted by J. F. Karcz, *Thoughts on the Grain Problem* in Soviet Studies 18, p. 414 (1967) from *Ekonomicheskoie obozrenie* 1929, no. 7, p. 190.

In 1924-25, in the wake of the 'scissors' crisis', the pre-war price relationship between crops and livestock products was fairly closely maintained; the main exception was oil-seeds which were proportionately much cheaper than grain and most other crops, perhaps as a result of the steep rise in the sown area and production of sunflower seed from 970,000 hectares in 1913 to 3.1m. in 1925, with a trebling of gross production.[25] From 1925 onwards the prices of livestock products climbed steadily and those of technical crops other than oil-seeds remained stationary, while oil-seeds began to catch up. Grain prices, however, remained unchanged in 1925-26, fell by 20 per cent in 1926-27, recovered slightly in the following year and rose dramatically in the last NEP year.

The movement of grain prices reflected a tenacious struggle

between the peasants and the Government on the grain market. At first the peasants tried to exploit the seasonal tightness of grain supplies in late winter and pushed grain prices up. The Government was embarrassed by the effect of this manoeuvre on the cost of living of the town population and the resulting dissatisfaction amongst the lower-paid workers and used administrative as well as economic measures to resist this trend and to maintain grain prices at the lower autumn level.

These measures proved effective, particularly after the elimination of private grain trade in the summer of 1926. Between October 1923 and March 1924 the index of agricultural wholesale prices had risen by 87 per cent with the approval of the Government as part of the liquidation of the 'scissors' crisis'; between October 1924 and March 1925, following a poor harvest, it rose again by 37 per cent and during the winter of 1925-26 it increased by 21 per cent despite an exceptionally good season and to the intense annoyance of the authorities. During the winter of 1926-27 the Government's counter-measures were successful and prices remained practically unchanged. The 15th Party Congress in December 1927 praised the 'considerable strengthening of the planning and regulatory rôle of the proletarian government in agriculture' which had just managed to stabilize bread prices in spring and autumn and had fixed agricultural prices 'in the interests of the national economy as a whole'.[26]

However, the bill for this successful operation was presented to the Government during the winter of 1927-28, when grain collection virtually broke down. In Stalin's words, the result was a recourse to 'emergency measures, administrative arbitrariness, violation of revolutionary laws, raids on peasant houses, illegal searches and so forth, which affected the political conditions of the country and created a menace to the *smychka** between the workers and the peasants'.[27] The law of dangerous over-success had operated with a vengeance and the NEP was to all intents and purposes dead.

* Union.

86

5. AGRICULTURE AND THE ECONOMIC
DILEMMA OF THE NEP

The threat of stagnation in agricultural output and the danger of a contraction in market supplies during the later 1920s confronted the Soviet power with an increasingly grave dilemma. Apart from the immediate consequences for the food and raw materials basis of the economy, it endangered the economic future of the economy, for the time was rapidly approaching when the Government had to call on the peasants for a large contribution towards the cost of financing the fundamental reconstruction of Soviet industry.

The quick economic recovery from the apparently complete ruin of industry at the end of the period of War Communism had taken place within the framework, and on the basis, of the fixed capital inherited from Tsarist Russia. Although in industry the limits to further expansion of output were less rigid than in grain production, and substantial amounts were invested in new equipment, the pace of growth slackened appreciably once the pre-war level of production was exceeded.

TABLE 11
Industrial gross production 1925-1928
(1913=100)

	1925	1926	1927	1928
All industry	73	98	111	132
Producer goods	80	113	128	155
Consumer goods	69	90	102	120
	% increase over previous year			
All industry	62	34	13	19
Producer goods	54	41	13	21
Consumer goods	68	30	13	17

Source: Narkhoz 1962, p. 117 and calculations therefrom.

In the literature of the time, the gradual fall in the rate of increase was known as the problem of the 'descending curve'; the extent of this phenomenon and the theoretical and practical implications of the problem were at first the subject of sharp public controversy and later on of the show trial of Mensheviks.[28] Although even towards the end of the NEP high growth rates were, in fact, achieved, the problem of re-equipping industry on a large scale and expanding its heavy industrial base assumed great urgency. The threat of stagnation in some of the most important consumer goods, such as the cotton industry, was a source of particular embarrassment because these products entered prominently into the exchange between town and village which was being slowed down by the endemic famine of manufactured consumer goods.

A substantial increase in the marketable surplus of agriculture was thus both the immediately most pressing need and the condition of solving the intractable long-term problem of the Soviet economy which depended on a massive increase in fixed investment, in agriculture as well as in industry. The question of how this increase could be obtained was at the back of much of the passionate and dramatic debate which took place throughout the final years of the NEP period and which made it 'intellectually the peak period of Soviet economic thinking'.[29]

The choices confronting the Soviet leadership after 1925 were so grave and so crucial for the régime that the emergence of opposite viewpoints was inevitable in any case, but the discussion was charged with explosive force through the political developments with which it was connected. This was the time of Stalin's inexorable rise to power through a process of shifting alliances within the Communist Party. The centre of the opposition consisted of Trotsky and his supporters who were later on joined, much to their misfortune, by Zinoviev and Kamenev after the latter had served their turn as Stalin's allies against Trotsky; finally, after the defeat of the 'united' opposition, Stalin established his full personal ascendancy by evicting his allies on the Right—Bukharin, Rykov and Tomski.

On the specific issue of how to increase the quantity of marketed agricultural produce there were two extreme points of view. One was represented, by and large, by the Trotskyists, who proposed that the peasants should be made to sell as much as possible by keeping the agricultural tax high (particularly on the wealthier, or rather less poverty-stricken, sections of the peasantry), market prices of agricultural produce low and industrial consumer goods prices high. The other policy was precisely the opposite: it was based on the acknowledged fact that it was the larger peasants who produced a higher proportion of their output for the market than the smaller peasants, who produced predominantly for their own subsistence. Hence they should be assisted to extend their output by removing, or at least relaxing, the restrictions of the 1922 Land Law on the leasing of land and the employment of hired agricultural labour.

In their most consistent and influential form, these alternatives were represented by Yevgeny Preobrazhenski and by Nikolai Bukharin who were, oddly enough, joint authors of one of the most popular expositions of Communist doctrine, *The ABC of Communism*. Preobrazhenski was a Trotskyist economist who developed his theory of the primitive (or, perhaps more correctly, the original) socialist accumulation of capital through the exploitation of the peasants as sellers of agricultural produce and buyers of manufactures; Bukharin, who had moved from the extreme Left to the extreme Right of the Communist spectrum, formulated his *credo* in the telling, but politically inept, advice to the peasants, *'enrichissez vous'*.[30]

Preobrazhenski's economic writings were perhaps the most thoughtful and successful attempt to analyse the economic problems of contemporary Russia within the framework of the Marxist scheme of the operation of the capitalist system as a whole. Given the 'economic and technological backwardness of the proletarian state, compared to the foremost capitalist countries'—which kept industrial prices in the Soviet market well above world prices—'economic equilibrium, which ensures expanded reproduction in the state sector, can exist only on the basis of non-equivalent exchange with the private sectors',[31] i.e, in practice with the peasants. Only through the planned expansion of industrial capacity, and in the first place the creation of an efficient engineering industry, would it be possible to maintain a balance between these unequal partners; otherwise, he predicted, 'the law of value will break through with elemental force into the sphere of regulating economic processes'.[32]

In less abstract terms he explained that, from the peasants' point of view, the only thing that mattered was 'cheaper industrial goods in the necessary amounts and appropriate quality. This economic contradiction turns into a social contradiction, into a growth of peasant dissatisfaction with the monopoly of foreign trade, into efforts to liquidate the forced attachment of the peasant market to Soviet industry, efforts to break through to the value relationships of the world market, to get rid of the multi-billion tax into the fund of primitive socialist accumulation'.[33]

Preobrazhenski's conclusion from this sophisticated analysis was far removed from the crude repressive policy against individual peasants which was actually put into effect a few years later. On the contrary, he regarded this clash of interest as 'a whip that impels the state economy to bring domestic industrial prices closer to world market prices. Rapid achievements in this way, accompanied by growth of state credit for organizing the production of middle, and particularly of poor, peasants and providing them with additional means of production, will weaken this social contradiction. Delays on this road will increase this contradiction, threatening to raise against the socialist sector in the first place, the capitalistically most developed elements of peasant agriculture and the corresponding strata of the peasant population which are most hindered, in their development along the bourgeois path, by the process of expanded socialist reproduction'; and, as if to dissociate himself in advance from the tragic stupidity of Stalin's collectivization policy, he spelt out his solution of 'the most fundamental question of the relationship between socialist development of the city and capitalist development of the village . . . a more rapid rate of socialist development will permit the endurance of a larger dose of capitalist development without great danger for the system as a whole'.

In practice, each of the alternatives proposed by the Left and by the Right of the Communist Party had its disadvantages: the first had been discredited in advance by the 'scissors' crisis' of 1923—as recently as 1960, with Stalinism officially out of favour, an official publication blamed the 'scissors' crisis' on the Trotskyist Y. L. Piatakov in his capacity as Deputy Chairman of the Supreme Council of the National Economy.[34] The lasting lesson of this crisis was the possibility that the ruthless exploitation of the monopoly position of the state-owned industry and its selling syndicates could provoke a buyers' strike by the peasants; thus it was as likely to lead to a curtailment of agricultural market production as to its expansion. The policy advocated by Bukharin and his associates was certainly going to strengthen the *kulaks* and to enable them to charge higher prices for their goods. As the *kulaks* were widely regarded as the breeding ground for a capitalist renaissance, this policy could be plausibly accused of favouring the transformation of the largely egalitarian post-revolutionary village into a potential counter-revolutionary force.

In the political discussions of the time these possible implications of the policies advocated by the right wing of the Communist Party received, of course, most attention. They were emphasized, and probably exaggerated, by the opposition and denied, or played down, by the official leadership which still consisted of a coalition between Stalin and the 'Right'. The choice between the Trotskyist programme of concentrating on the building of a heavy industrial base and the Bukharinist policy (which was supported by many economists outside the Communist ranks) had to be made on political grounds. The demand for a stepping up of industrial investment at the expense of the peasants involved acute friction with the great majority of the rural population and perhaps open conflict; the strength of the Trotskyist position lay in its appreciation of the inevitability of such a conflict. The weakness of the alternative policy was precisely its lack of political perception for the consequences of the economic measures it advocated with a great deal of persuasive force. Nothing could be more sensible than to call for more investment in agriculture and the consumption goods industries, but this could only be achieved by permitting the peasants to earn enough from the market to increase both their personal consumption and their investment. However, the inevitable corollary to this desirable development would have been to make the growth of industry dependent on its own resources, which were pathetically unequal to this task, and thus to increase the dependence of the Government on the peasants.

The policy actually pursued by the leadership under the pressure of events and of the conflicting forces within the ruling party managed to combine the worst of both worlds. The influence of the Right was discernible in the encouragement of the larger peasants through changes in the Land Law, but the price policy pursued towards the peasants made it less and less attractive for them to increase their marketable surplus. The result was total deadlock which was broken rather than overcome by the use of force.

There is no more fitting epilogue to this tragic history than the striking phrases in which Preobrazhenski, well before the event, characterized 'the forced character of our co-operation with private economy. There is co-operation in prison, too. Are we not in a sort of concentration camp along with the capitalist elements of our economy? We are at one and the same time warders and prisoners.'[35] It is one of the supreme ironies of Soviet history that the failure of the Communists as a party, and particularly of their leadership, to grasp the implications of this profound insight transformed the Soviet Union for years in good earnest into a giant concentration camp.

VI

The Second Agrarian Revolution and its Aftermath

1. THE WAR AGAINST THE PEASANTS

Industrial Needs and Agricultural Resources

The 'contradictions' confronting Soviet society at the end of the NEP era can be summed up in a series of simple and forbidding propositions: no chance of economic reconstruction and military security without a heavy-industrial base; no heavy industry without massive investment financed from outside industry, in the last resort through the agricultural surplus product; no growth in the marketable surplus of agriculture without either a rise in farm output or a fall in farm consumption, or both; no substantial rise in farm output without heavy investment in industrial producer goods, and no voluntary cut in farm consumption of agricultural produce except in exchange for manufactured consumer goods; no increased production of manufactures for the village without prior massive investment in the creation of a heavy-industrial base.

The vicious circle was apparently complete and the chances of breaking out from its deadly embrace were, therefore, correspondingly remote. Well might the social-democratic critics of the Communist régime, such as the veteran Karl Kautsky, triumphantly denounce the deadlock reached after more than a decade of Communist rule,[1] for the unsatisfied needs of Soviet society surrounded the Bolsheviks like a wall of negatives. It needed boundless ruthlessness and the pressure of an increasingly more explosive internal situation to escape them through a head-on collision.

This point was reached in the early months of 1928, when the dissatisfaction of the peasants with the terms of exchange between agricultural and industrial goods dangerously reduced the quantity of grain coming on the market. As it happened, an

93

increase in grain deliveries had become imperative at this very moment, not only in order to meet the growing needs of the expanding towns but also in order to pay for the large imports of industrial plant required for the forced industrialization policy on which the régime had at last decided. The description of the resulting situation as 'the bread front' was no exaggeration, and within months open warfare erupted between the State and a large proportion of the peasantry.

When Trotsky and his supporters advocated the adoption of a policy of forced industrialization in 1925, there was still a fair amount of slack in most branches of Soviet industry which could have been taken up without the need for *prior* large-scale expansion of fixed investment; hence there would have been at least some chance of coaxing the peasants along this path by increasing the supply of manufactures in exchange for greater supplies of agricultural products which could be exported to pay for industrial equipment from abroad. When the Soviet leaders actually embarked on the policy of forced industrialization in 1928, agricultural expansion had virtually ceased, and with it the growth in the output of some of the most important consumer goods industries. The question of how to stimulate the supply of the desperately needed additional agricultural market produce had become incapable of a rational answer.[2]*

The Communist régime was compelled to raise this question, nevertheless, with the utmost vigour, for industrialization based on coal, oil and steel was both time-consuming and enormously expensive in scarce resources. It involved the sinking of coal-mines and oil-wells, the building of steel-mills and engineering works and the purchase of expensive foreign machinery for the

---

* In defence of the official Soviet policy against this criticism, Mr Maurice Dobb claimed that 'it does not follow that what may have been practicable in 1928 or 1929 was necessarily practicable at an earlier date when both industry and agriculture were weaker or that what circumstances may have made imperative in 1928 was demanded also at a time when the *kulak* influence over the village and the grain market was smaller'.[3] This statement assumes, against the evidence, that industry and agriculture were 'stronger' when the opportunities for expansion at low marginal cost had been exhausted and that the disastrous policy made 'imperative' by the emergency conditions of 1928 would also have been inevitable, even if the emergency had been prevented by purposeful action instead of being allowed to occur by the bureaucratic inertia responsible for the official refusal to act in time.[4]

purpose of making in due course similar machinery in the new works. Meanwhile, large funds were needed to maintain the construction workers while they were building the factories, to pay for machinery imports and the high fees needed to attract foreign specialists for training Russian managers and workers in the use of the new techniques.

In the final analysis, the peasants had to supply the foodstuffs and raw materials needed for all these purposes and to do this without payment of the only kind which was acceptable to them—useful manufactured goods. This was not the situation envisaged by Preobrazhenski's sophisticated theory of primitive socialist accumulation through weighting the terms of trade against the village, but barefaced expropriation of the *muzhik* on a scale hardly ever envisaged before.

Soviet agricultural policy during the ensuing years can be properly understood only against this background. It was essentially a power struggle aimed at subjecting the peasants as producers and as suppliers directly to the authorities by relegating the operation of market forces to the fringe of the economic process.

Collectivization in Theory and in Practice

The technical superiority of large-scale crop cultivation and hence its ultimate triumph over small subsistence farming, and even more over the dwarf holdings characteristic of so much of Russian agriculture, had been an article of faith for socialists of various persuasions for generations; the Communists in general, and Lenin in particular, had no doubt about its truth.

In its most absolute form, the proposition is by no means self-evident, for large-scale production frequently yields higher returns per man than family farming but lower returns per acre. As in peasant economies land, at least readily cultivable land not in need of large-scale investment through irrigation, fertilizers, etc, is generally much scarcer than labour, this difference may well tilt the balance of advantage in favour of small-scale farming until the development of industry has made labour scarcer and investment resources more plentiful.

In Russia, however, the land use of most small farms was subject to traditional handicaps which went far to offset this factor.

The tiny fields on which the bulk of the crops was grown were extremely wasteful of land as well as of labour, mainly because of the subdivision into strips and the adherence to traditional and economically obsolete forms of crop rotation. Plot size and configuration naturally also prevented the application of mechanized cultivation and, therefore, a quick improvement in the productivity of labour. However, this was not yet a major drawback in the conditions of the late 1920s, when the towns were unable to absorb the surplus labour thrown up by the village. The real priorities were a bigger sown area, which could be obtained most economically by better use of available land and higher yields through more advanced technological methods, above all through improved seeds and animal breeds.

During the NEP years the dissemination of agronomic knowledge had increased, and at least part of the improvement was attributable to the State farms ('sovkhozes')[5] which numbered 4,651 with 3.2m. hectares in 1925-26 but only 1,407[6] with a sown area of 1.1m. hectares[7] in 1928. However, in the new era the strategic use of the State farms was not that of agricultural model establishments but that of an alternative to the traditional forms of farming on land which had been left uncultivated in the past, mainly because of drought hazards. The creation of gigantic 'grain factories' in such areas under direct Government control was regarded as one way of ending the dependence of the régime on the variable surplus product of some twenty-five million peasant households.

The main effort of putting the relations between the Communists and the peasants on a new basis was, however, made in the villages. It was, again, generally accepted socialist theory that this involved the replacement of the tiny individual production units by a network of agricultural co-operatives. The three traditional forms of co-operative or 'collective' farming during the 1920s were the *communes,* which were a thorough-going and unsuccessful experiment in communal working and living, the agricultural *artels* or collective farms ('kolkhozes') in the later most usual sense which worked the land in common but maintained individual households as centres of consumption and of subsidiary production, and the much looser joint partnerships (*tozy*).

96

In 1925 collective farms of all types numbered 21,923 with over one million participants, almost 300,000 homesteads and about 3¼m. hectares of agricultural land;[8] by the end of 1928 their number had increased to 33,000 with about 400,000 holdings and an average of 96 hectares of land per collective farm;[9] 5.4 per cent were *communes,* 38.4 per cent *artels* and the majority were *tozy*.[10]

Given this diminutive scale of the public sector, the transformation of the individualistic village into a system of large producer co-operatives was universally regarded as a task whose completion was bound to take many years. In December 1927 the 15th Congress of the Communist Party laid down 'directives' for the agricultural policy of the Five Year Plan in an inordinately long resolution on *Work in the Village* and the main Congress resolution was quite explicit on the fact that the fruits of the encouragement of co-operation would mature only very slowly: 'One of the fundamental conditions of the evolution of the Soviet Union towards socialism appears to be the raising of the productive forces of the village and the growth of the prosperity of the broad peasant masses. The socialist town can draw the village towards itself only on this path, by supporting in every way the gradual transition to collective forms of the individual property-owning peasantry, which will be for a considerable time the basis of agriculture as a whole.'[11]

These instructions were embodied by the planners in quantitative targets for the share of the public sector at the end of the Five Year Plan both in its standard variant and its optimum variant:

TABLE 12
Percentage share of the public sector in agriculture: 1927-28 and 1932-33

| | 1927-28 (actual) | | | 1932-33 (planned) | | | | | |
| | | | | Standard variant | | | Optimum variant | | |
	State farms	Coll. farms	Total	State farms	Coll. farms	Total	State farms	Coll. farms	Total
Gross production	1.2	0.6	1.8	2.9	10.1	13.0	3.1	14.0	17.1
ditto, grain	1.1	1.0	2.1	3.8	10.8	14.6	4.3	15.5	19.8
Market production	3.6	0.8	4.4	8.0	14.9	22.9	8.6	16.7	25.3
ditto, grain	3.7	3.8	7.5	16.4	22.6	39.0	17.3	25.3	42.6
Sown area, total	1.1	0.9	2.0	3.2	10.3	13.5	3.5	14.3	17.8
ditto, grain	1.1	0.9	2.0	3.0	9.5	12.5	3.3	13.1	16.4

Source: *Piatiletni Plan,* I, 90.

To increase the share of the public sector in agricultural gross production in five years from a mere 1.8 per cent to 13.0-17.1 per cent of an expanding total would have been an enormous feat by any standard. For grain, the basic commodity, the expansion was to be even greater, particularly for market production on which the success of the First Five Year Plan ultimately depended. Nevertheless, the planners expected that in the most favourable circumstances the public sector would provide less than half the total at the end of the period, and despite the great hopes put into the new State farms, much the larger part of the increase was expected to come from the collective farms.

The First Five Year Plan was an impressive, though far from flawless, planning exercise and at the same time an instrument of mass propaganda on a gigantic scale. It is a moot point whether the anticipated rise in the market production of grain was at any time a rational forecast, but it was abandoned with the decision to 'over-fulfil' the collectivization programme to a tragic extent. The history of this struggle need not be told in detail in this context, but its main results were an essential element in the relations between the Soviet power and the peasants and are even now of more than historical importance.

The starting point of the conflict was not the need, however genuine, to overcome the backwardness of the post-revolutionary Russian village nor the political danger threatening from the counter-revolutionary activities, real or alleged, of the *kulaks*. It was the opposition between the Government and the towns as bread consumers and the peasants as food suppliers which found its expression in the 'bread front'. The policy followed by the régime during the succeeding years can only be judged by military standards. Like other wars it involved enormous wanton destruction and left victors and vanquished impoverished in important respects; it decided nothing except the one question which of the contending parties was stronger than the other.

In the history of civil wars the collectivization struggle of 1929-32 occupies a position of unenviable eminence. It was less ferocious but more prolonged and systematic than the fierce peasant wars against the state which were such a commonplace of Tsarist Russia, but it was no less ruthless a crusade than the Franco-Papal war against the Albigenses. It achieved its ostensible purpose only by sacrificing the real aims which could have justified it. In its process the balance of forces within Soviet society and the very character of the régime underwent changes which put off political progress for twenty years and left deep scars which have even now barely healed.

The great debate within the Communist Party which preceded the collectivization crisis, was, perhaps, the most intense and searching preparation for a conscious new departure in economic policy which has ever taken place in such circumstances, but the policy actually pursued was primitive, callous and wasteful in the extreme. It is idle to speculate on what might have happened if either the Left or the Right had gained the upper hand, but in the event the battle was won—and policy decided—by the 'general line' represented by the General Secretary of the Party whose contribution to the debate had been secondary but whose ascendancy at its end was absolute.

The issue was not decided by the force of arguments but by the party machine which found its perfect representative, and its master, in Stalin; soon afterwards it went into action to solve the great dilemma of the position of the peasantry within Soviet society by the truly bureaucratic means of regimentation, intimidation and brute force. By such measures the number of collective farms was doubled between June 1928 and October 1929, when the campaign was intensified in connection with the stepping up of hostilities against the *kulaks,* and the percentage of collectivized households doubled again in the last quarter of 1929 when it approached 5m. The tempo was stepped up even further in January and Februrary 1930 and at the beginning of March 1930 collectivization was said to have extended to 59.3 per cent of all peasant households or roughly 15m.[12] As far as such claims had any meaning, they presupposed a state of open warfare between the State and the peasants: 'Indeed the events of this period were marked by a wave of disorder, violence, looting and debauchery which swept the whole country.'[13]

Millions of peasants retaliated partly by sporadic acts of violence but mainly through killing off their livestock and the situation reached such a pitch that Stalin himself, though mainly responsible for this policy, found it necessary to apply the brake and to reproach the allegedly over-zealous party machine for its 'excesses'. The number of collective farms existing on paper fell sharply during the spring and summer of 1930, but collectivization took a further leap forward between 1930 and 1931, with the number of collective farms rising from 85,950 to no less than 211,100.[14] From this level it increased only very slowly until the outbreak of the Second World War. The number of collectivized households rocketed from 400,000 in 1928 to 14.7m. in 1932, 18.1m. in 1937 and 18.7m. in 1940.[15] The share of agricultural land held by individual peasants declined from 96 per cent in 1928 to less than 10 per cent in 1940, including the newly-acquired Baltic republics where collectivization was carried out only after the war. However, the subsidiary private plots of collective farm members and of State farm and other workers and employees accounted for another $3\frac{1}{2}$ per cent.[16]

When the First Five Year Plan was officially completed at the end of 1932, the transformation of the Soviet countryside from a stronghold of the individual, independent peasant into a network of State and collective farms under the control of the authorities had been, by and large, achieved. The next few years witnessed a certain consolidation of the new system which was formally confirmed by the Second Collective Farm Congress in 1935 which gave the collective farms their land 'in perpetuity' —a gesture of no practical importance but a token of the official determination to round off the second agrarian revolution by a tolerable compromise.

The Decline in Output and its Effects

The failure to solve an economic problem of great difficulty and delicacy while there was still time for the use of economic measures transformed it into a political issue of the most imperative, and therefore of the most primitive, kind: a trial of strength which degenerated into a naked struggle for survival with no holds barred. In this conflict the need to expand agricultural production as a condition of increasing the marketable surplus of agricultural products was lost sight of and the dominant consideration became the control of whatever output could be obtained. The public investment in agriculture was concentrated on mechanization of the main processes of crop production—mainly the sowing and harvesting of grain—while at the same time the excesses of the fight against the *kulaks*, which was often a term used to cover the majority of the peasants, led to the destruction of existing resources, particularly of livestock, on a gigantic scale.

The years of the First Five Year Plan were, therefore, on balance a period of declining agricultural gross production, though there were some relatively bright spots in the picture. Apart from the very erratic sugar-beet crop, the so-called 'technical crops' and particularly textile fibres did relatively well, but the gross production of grain and potatoes actually declined during the last two years of the period below their level at the end of the NEP (Appendix, Table II).

The food demand of the towns and grain export commitments had to be met in order to keep the industrialization programme going and the quantity of grain left to feed the peasants and their livestock was drastically reduced. According to recent Soviet publications the grain available for human food in 1932 was 35.2m. tons compared with 36.6m. tons in 1928, while the tonnage available for feed declined from 20m. tons in 1928 to only 13.8m. tons.[17] One of the consequences of this catastrophic situation was a sharp drop in grain exports compared with their level in 1930 and 1931, but, in addition, living standards had to be cut drastically, particularly in the village, where Government 'procurements' absorbed over one-third of the gross grain harvest. (In 1932 no less than 36.9 per cent.[18])

Although scenes of violence approaching in some cases to civil war were by no means infrequent, the main weapon of the peasants in their struggle with the authorities was passive resistance, mainly in the form of sullen refusal to work effectively on the newly-formed collective farms and to hand over their means of production, above all their livestock, to them. The failure of the grain production plans would have in any case prevented the maintenance of herds and flocks at their previous levels, but the fall in livestock numbers was already well under way before the disastrous seasons of 1931 and 1932. Rather than giving up their horses and cattle, sheep and pigs to an instrument of Goverment policy, many peasants slaughtered their livestock, even though they frequently were not even allowed to consume the meat themselves. In the single year of 1930, with an unusually good harvest, the stock of horses fell by 4 million, that of cattle by over 8 million, pigs by 2½ million and sheep by over 25 million (Appendix, Table IV). In 1932 the number of horses and pigs was only about one-half of the pre-collectivization level, that of cattle less than three-fifths and that of sheep and goats little more than one-third. The catastrophic fall in livestock numbers was, in the first place, caused by the policy of the authorities and only later on by the dire feed shortage which was, of course, indirectly also the result of the second agrarian revolution. In 1928, the official index of livestock production was 137 per cent of the pre-war level; four years later it had dropped to only 75 per cent. By 1932 milk production was more than a third lower than at the beginning of the 'plan period' and meat had fallen by over 40 per cent, eggs and wool by about 60 per cent. (Appendix, Tables I and III.)

The effect of the calamitous drop in the production of most foodstuffs was a severe decline in mass living standards. The promises of the First Five Year Plan in this field had never been seriously-based realistic forecasts but little more than propaganda; they assumed increases in productivity which could not have been achieved even in favourable conditions and which proved in the end totally illusory.

TABLE 13

Per head food consumption in 1927-28 and planned consumption in 1932-33

Item	unit	1927-28		1932-33	
		Actual consumption		Planned minimum	
		Urban	Rural	Urban	Rural
Bread*	kg	179	221	183	238
Meat	kg	49.1	22.6	56.0	24.7
Milk products†	kg	218.0	183.0	301.0	207.3
Eggs	pieces	90.7	49.6	138.7	69.5

* Grain equivalent.   † Milk equivalent.
Source: *Piatiletni Plan*, I, 106.

The trend of total output and of population makes it obvious that consumption per head must have fallen sharply, in some cases by more than half. Though later data for 1928 differ slightly from those on which the previous table is based, they indicate that the rural population had to accept a severe cut in the consumption of all types of food (except potatoes), while the non-agricultural population had to make do with more bread and potatoes and very much less meat and butter.

TABLE 14

Annual norms of consumption of agricultural products (in kg)

	Agricultural population					Non-agricultural population				
	1928	1929	1930	1931	1932	1928	1929	1930	1931	1932
Bread*	250	245	241	234	215	174	170	198	208	211
Potatoes	141	146	147	145	125	88	109	136	144	110
Meat and fat	24.8	29.5	21.2	20.0	11.2	51.7	47.5	33.2	27.3	16.9
Butter	1.6	1.2	1.0	0.8	0.7	3.0	2.8	2.3	1.8	1.8

*Grain equivalent.

Source: Y. A. Moshkov, *Zernovaya problema v gody sploshnoi kollektivisatsii selskovo khoziaistva* SSSR (1966), p. 136 (slightly abbreviated).

The struggle between the régime and the peasants reached its grizzly climax in the winter of 1932-33, when a man-made famine swept parts of the Ukraine, the northern Caucasus and Kazakhstan; the number of its victims will probably never be known but it has been estimated at four to five million people.[19]

The consequences of the fall in agricultural production during the agrarian revolution of 1929-32 were not limited to drastically-reduced food supplies and severe local famine. Reductions of this order of magnitude could not be accommodated by a general tightening of belts without affecting the working efficiency and the political loyalty of the key sectors of the population, the inadequate 'cadre' of skilled industrial workers and the directing and repressive élite of the Communist Party. The fall in food supplies was, therefore, not spread evenly over the population but was accompanied by the emergence of an increasingly sharp social stratification.

Though the privileges granted to a minority were modest in themselves, their social effects were intensified by the incredibly low level of real living standards of the unprivileged majority.

At the time these privileges were not expressed to any great extent in cash terms, because the dislocation of markets made sale and purchase comparatively unimportant methods of procurement, but in the direct segregation of different groups of consumers. The majority had to rely on the basic rations available, if at all, in the general State and co-operative stores; the minority were permitted to shop in the special distribution departments attached to enterprises of varying degrees of importance serving exclusively their élite customers, with the Kremlin and the GPU at the top of the hierarchy. The effects of this development on the structure and policy of the Soviet régime were to outlast by a long time the immediate emergency which had given rise to these measures.[20]

2. THE COMPROMISE SETTLEMENT

Neither the political foresight of the rulers nor the enthusiasm of the peasants for the new system of compulsory co-operation played any part in creating the conditions for a settlement of the acute conflict between the Soviet power and the peasants; a compromise was forced on both parties by dire necessity.

The peasants discovered by bitter experience that they could not shake the determination of the Government to obtain a direct and controlling interest in the way they tilled the land, raised livestock and disposed of their produce. They had to accept the fact that everybody who dared openly to oppose the official line was crushed by brute force, and that millions were allowed to starve if their output had fallen, through passive resistance or the vagaries of the seasons, to a level below that required to keep them alive after the Government had taken its share of the harvest.

The Communists, for their part, had to acknowledge that failure to make the new agricultural system at least bearable to the mass of the peasantry condemned the whole country, including themselves, to hunger and poverty, and that the huge losses inflicted on the agricultural economy during 1929-32 could only be made good if the peasants regarded it as worth their while to produce more.

Although the situation towards the end of the First Five Year Plan thus resembled that of 1921 in some respects, the differences were more marked than the similarities. The great increase in the economic and administrative strength of the régime was plain for all to see. Heavy industry was developing very fast, though more slowly than the huge sacrifices made by the population would have justified, and the reorganization of agriculture freed the Government from its near-complete dependence on the goodwill of the peasants. There never was any question of retreating from the main tenet of a large-scale agriculture under the control of the Government, but ways had to be found of combining it with some inducement to the peasants; only then would they be willing to supplement the stark emergency rations to which the towns had been reduced by the winter of 1932 with enough marketable produce for the modest creature comforts which the town population had experienced during the later years of the New Economic Policy.

The 'Public Sector' in Agriculture

In the broad concept of Communist policy, State farms and collective farms were complementary to each other: the former were to be specialized agricultural factories under direct Government control, particularly useful for expanding grain cultivation in new areas where climatic hazards had previously prevented permanent agricultural settlement; the latter were to bring the amorphous mass of almost twenty-five million peasant holdings under the indirect but effective sway of the authorities.

The exaggerated hopes put at first in large-scale State farms as a source of marketable grain led to a reaction in the later 1930s and to a considerable fall in their sown areas, with the emphasis switched to animal husbandry rather than grain production. In 1940 there were altogether 4,159 of them—about half as many again as in 1930. The number specializing in field crops had risen relatively little (from 639 to 784) and those specializing in orchards had gone up only from 402 to 494. The main concentration of effort was on the side of producing livestock and livestock products, where the number of State farms increased from 1,077 to 2,035 (meat and milk producing farms doubled from 434 to 870 and piggeries from 350 to 666).[21]

Despite the retreat from the early gigantomania, the average State farm in 1940 remained a very large enterprise, with an area of 12,200 hectares of agricultural land of which 2,800 were under crops, 285 workers and 24 tractors (in 15-h.p. units).[22] State farms occupied 7.7 per cent of the total sown area and 6.9 per cent of the grain acreage.[23] They supplied the Government with 3.67m. tons of grain, slightly less than 10 per cent of total grain procurements and a considerably higher proportion of livestock products, including 338,000 tons of meat and poultry (in terms of live-weight), compared with 881,000 tons from the collective farms, 1.01m. tons of milk and milk products (3.25m. tons from the collective farms) and 85m. eggs (100m. from the collective farms.)[24] They played a significant part in public agriculture, but their importance was distinctly subsidiary to that of the collective farms, particularly in the key sector of grain production for the market.

The collective farms numbered in 1940 no fewer than 235,500. They included 18.7m. peasant households and had a sown area of 117.7m. hectares of which 91m. were under grains[25]—about four-fifths of the total for all branches of agriculture. They were the predominant source of the public trade system for all crops, but their importance as a source of livestock products was limited, partly because of the stronger position of the State farms in this branch of agriculture, but mainly because of the strength of the peasants as individual producers in this field.

The scale of operations of the collective farms depended so much on regional or even local conditions that national averages convey only a very imperfect view of the situation. At the end of 1932 the average collective farm contained 71 households with 434 hectares of commonly-held land under crops, 42 head of cattle (including 13 cows), 15 pigs and 54 sheep and goats.[26] In the following year, exactly one-third of them contained fewer than 30 peasant households, and three out of five had fewer than 60; at the other end of the scale, one collective farm in five had between 100 and 300 households and one in thirty more than 300. The size distribution according to the acreage under crops shows a similar dispersion, with almost half the collective farms having less than 200 hectares sown area and only one in eight over 1,000.[27] The majority of collective farms were, therefore, modest agglomerations of very small holdings, others were substantial large-scale enterprises.

After the first few years of hectic and disastrous experimentation, the authorities relied for the control of the collective farms on two key instruments: the Machine and Tractor Stations (MTS) and the Government bureaucracy under the tutelage of the Communist Party. At the end of 1940 there were 7,069 MTS and they deployed the main resources of agricultural machinery; their staff totalled 537,000—an average of about 75 per unit—and they disposed of 435,000 tractors (equal to 557,000 in 15-h.p. units) and 135,000 grain combine harvesters.[28] The collective farms were obliged to make use of the MTS, where their services were available, in exchange for a certain share in the harvest, and this arrangement ensured maximum utilization of scarce machinery and part of the supplies needed for the urban population. The everyday control of agricultural production remained difficult even after the concentration of 25m. independent peasants into fewer than a quarter of a million collective farms, and the relatively small number of the MTS (about one for thirty-five collective farms) gave the authorities improved leverage for this purpose.

The tasks of the MTS were not confined to the technical management of mechanized land cultivation (ploughing, sowing and harvesting): 'They help the collective farms set up plans of production and finance. They fix the correct crop-rotation system, assist in the organization of work and the allocation of income, the training of leaders, the setting up of accounting systems, the organization of competition, and the struggle to increase soil productivity.'[29] They were, in fact, the main instrument for the transmission and execution of cropping policy in the large majority of collective farms and for this reason they had a strong leavening of Communist Party members amongst their personnel. For a time party control was embodied in special 'political departments', but it remained important even after their abolition at the time of the general settlement of 1935 which gave the collective farms a wider measure of autonomy than before.

107

The collective farms were also supervised by the local organs of the Government, particularly the district (*raion*) and regional (*oblast*) Soviet Executive Committee and the local organs of the Ministry of Agriculture. The latter was the technical authority for all matters not covered by the MTS, and particularly animal husbandry. The District Executive Committee was responsible for allocating, through the competent technical body, the local shares in the annual production and procurement plans of individual agricultural products to the collective farms in its area. The Party was charged with ensuring that official policy was carried out by the bureaucracy and by the management of the collective farms.

In order to make the system work in practice, central direction from above had to be combined with a certain measure of co-operation from below, for agriculture needs everywhere a certain degree of practical initiative at farm level. This crucial problem was solved only very imperfectly, though a tolerable compromise was arrived at by a combination of organizational and economic methods. By and large, the managers of the collective farm enterprises were given a modest measure of autonomy and these enterprises were, to some extent, limited to the main agricultural activities characteristic of their economic region. This involved in practice the recognition of a fairly wide sphere of individual interest for the collective farmers.

As in other areas of Soviet society, spontaneous activity from below had to be fitted into a framework of bureaucratic regulation and was, therefore, severely limited in practice. This was particularly noticeable in the position of the collective farm chairman, who was formally elected by all the members of the collective farm but was either selected by the real rulers of the district—the Party secretary and the boss of the District Executive Committee of the Soviet—or had at least to be acceptable to them. (An unusually frank acknowledgement of the continuation of this situation as late as 1962 was given by Khrushchev in his Report to the Central Committee of the Communist Party in March 1962.[30]) In the early stages of collectivization Communists had simply been sent from the towns to act as chairmen, because political reliability was regarded as infinitely more important than technical competence. In such conditions elections

108

were a mere formality, but later on they may have ensured that the authorities had to pay some heed to the feelings of the peasantry in choosing suitable candidates for this key position.

According to the Model Statute of 1935 the chairman was not much more than the presiding officer of the board of management and the membership meeting, but in fact he was (and remains) the responsible manager who has to account for the performance of the farm to the authorities; he 'could only manoeuvre between the Party organs and those of the Commissariat for Agriculture, "registering" the discontent of his members and pleading their needs. Any expression of opposition immediately resulted in removal from the chairmanship by joint action of Party and State organs'.[31]

The chairman has the assistance of the management board and of specialists (including the farm accountant) as well as of the 'brigadiers', each of whom is in charge of a 'brigade' or special work unit, whose job may be the cultivation of certain crops or the care of livestock.

The main assets of the collective farm were initially the land of its members and certain means of production, above all the livestock which the peasants had been induced or compelled to hand over. This 'indivisible fund' was defined by the First All-Union Congress of Agricultural Collectives in 1928 as not subject to division amongst the members of the co-operative and liable to be handed over to its successors in case of liquidation of the collective farm.

During the First Five Year Plan (1928-32) allocations to this fund amounted to 3,000m. rubles (post-war, pre-1961) and as the total balance in the 'indivisible fund' amounted to 4,700m. at the end of 1932,[32] it could be concluded that the assets originally brought into the collective farms were valued in their books at 1,700m. rubles. From 1933 until the outbreak of the Second World War in 1941 the collective farms invested out of their own resources another 22,800m. rubles. For the whole period from 1928 to 1941 their own investments were almost as large as those of the Government in the whole of agriculture, and from 1938 onwards they were considerably larger.[33]

The hold of the State on the output of the collective farms was no less powerful than its sway over their assets. Apart from the share of the crops taken in kind by the MTS, the Government determined the delivery quota of the various products to be supplied by the farms and the prices at which they had to be handed over; these were low from the start and became almost nominal with the growing depreciation of the currency. In addition, supplementary procurement at somewhat higher but still grossly inadequate prices was prescribed and only after all obligations to the Government had been discharged was the collective farm allowed to sell any surplus on the 'collective farm market'.

The distribution of the collective farm income depended to a large extent on the price formation which determined, in the last analysis, how much would be left over for distribution. In theory, the collective farmer was a partner in a co-operative enterprise and not a worker like the State farm staff. He was, therefore, not entitled to a firm wage or salary but to a proportion of the net income produced by the collective farm. This meant in practice that he was the residuary claimant on a fund which had been seriously denuded by a more powerful party: the State in a variety of rôles. Public trade supplied a minimum of industrial producer goods at high prices, the tax-gatherer collected income-tax and the local bosses insisted on a stiff allocation of resources to the indivisible fund, partly to finance expansion and partly to supply social and cultural needs.

After current expenses, income-tax and investment came the collective farm members. In the early days of the collective farms, income was frequently distributed on a population basis or according to the number of workers, but with the progress of mass collectivization the less egalitarian principle of labour days, in use in some collective farms from the beginning, had to be generally applied. In February 1935 a directive of the Ministry of Agriculture laid down payment according to the number of 'labour days' rendered by a collective farmer. These evaluated the time spent on different jobs or by different types of workers according to a scale within seven grades; the lowest day's work was equal to half a labour day and the highest—that of the chairman of large collective farms and the most difficult mechanical jobs—was worth two labour days.[31]

However, the value of a work day was not fixed: it depended on the income of the collective farm less all outgoings, including investment, and was ascertained only after the end of the year both in cash and in kind. Thus it depended partly on the prices paid by the Government and those obtained in the collective farm market, partly on yields and partly on the efficiency of farm labour and management.

As in industry, during the later 1930s the dominant official method for raising the productivity of the collective farms became the appeal to the self-interest of the individual and, above all, to that of all levels of management. The process began with the MTS and was gradually extended to the collective farms, starting with their chairmen and brigadiers. It reached its climax in a decree of December 10, 1940, which promised substantial premium payments for the over-fulfilment of plans. Individuals and groups mainly responsible for extra output were allowed to keep part of the excess in kind and special promises were held out to the managerial staff. The chairman was assured of a fixed salary, plus certain labour day credits, plus bonuses rising with the extent of over-fulfilment of production plans; specialists were to receive 70 per cent of the chairman's bonuses, brigade leaders, managers of livestock farms and, to a lesser extent, team leaders, were all to receive bonuses at higher rates than their subordinates.[35]

These appeals to material self-interest operated within a system based, in the last resort, on the intense exploitation of the peasants. Their aim was not to offer the peasants a greater share in the produce of their own work but to redistribute the share which the régime decided to leave them in favour of the more highly qualified and better placed, and therefore at the expense of the ordinary collective farm workers. It is, therefore, doubtful whether these efforts would have been successful, even if the war had not intervened. In the final analysis, they would necessarily have reduced the interest of the ordinary *muzhik* in the collective farm even further, and correspondingly increased his concentration on his subsidiary individual plot.

The legalization of this plot had been an essential part of the compromise between the régime and the peasants which terminated the second agrarian revolution. The official policy towards the private economy of the collective farmers and others was a kind of fever chart of the new agricultural system.

Collective and Private Interests

The main outlines of the settlement of the collectivization crisis began to emerge very early in the course of the conflict. In March 1930 Stalin was forced to retreat temporarily and blamed the excesses of the bureaucracy which he had unleashed under the slogan of 'liquidating the *kulaks* as a class'; at the same time he also defined clearly the type of collective farm to be set up— the *artel* or production co-operative: 'In the agricultural *artel* the most important means of production, mainly in the cultivation of grain, are socialized: labour, use of land, machinery and implements, working stock and farm buildings. Small vegetable gardens and orchards, however, houses, a certain proportion of the cows, poultry and small livestock, etc, are not socialized.'[36] Though at the time this precept was widely disregarded, it formed the basis, or at least an essential element, of the compromise between the régime and the peasants.

The bulk of the land was to be transferred to the main branches of public agriculture, the State farms and the collective farms, and on this cardinal point there was to be no wavering. In the last year of the First Five Year Plan, 1932, the collective farms contained 91.5m. ha. of a total sown area of about 130m. By 1937 their lands had grown to 116m. ha. and increased further, though slightly, to 117.7m. in 1940; with State farms and other public farm enterprises accounting for over 13m. ha., only 4.5m. ha. were in the hands of the collective farmers and a further 0.8m. in those of workers and employees in 1940 as 'subsidiary personal plots'.

By 1939, individual peasants had virtually ceased to exist, though during the later 1930s their submission had been brought about more by financial pressure (through heavy taxation) than by administrative action. However, between 1939 and 1940 the total sown area in the country rose by 16.7m. ha. through the incorporation of the Baltic republics, Moldavia, etc, from 133.7m. to 150.4m. ha., and the bulk of the new lands remained untouched by collectivization until after the Second World War; hence in 1940 over 14.1m. ha. were still in the hands of individual peasants.[37]

The compromise settlement was codified in the Model Statute (or charter) of 1935 which was issued in connection with the

Second Congress of Collective Farmers. The collective farms were assured of the use of their lands 'in perpetuity', and this fundamental right was confirmed by Article 8 of the Constitution of 1936, but this did not prevent the massive amalgamation of collective farms after 1950 nor their conversion into State farms after 1956. Though the size of the 'subsidiary personal plots' or allotments varied to some extent with the political situation, it was generally one-quarter to half an ha. They were generally situated in the immediate vicinity of the peasants' homes and used mainly for growing potatoes, vegetables and fruit. The land on which the staple products of Soviet agriculture were cultivated—grain, cotton, sugar beet, sunflower seed, etc—as well as the bulk of grazing land formed part of public agriculture.[38]

A much less uneven division took place in animal husbandry. Towards the end of the NEP period, the average livestock holdings per farm had been pitifully small—a single horse, cow, calf or ox, and pig, a few sheep or goats, chickens and geese; their distribution, however, had not been fully egalitarian, with perhaps a third of the peasants owning not a single horse and a quarter not even owning one cow, while others may have possessed two, three or more. Collectivization disposed of these 'kulaks' and public agriculture appropriated the bulk of the radically reduced work stock: in 1939 collective farms, State farms, etc, owned 15.6m. out of 17.2m. horses or over 90 per cent of the total, compared with 36.1m. mainly owned by individual peasants in 1928.

The situation for productive livestock was very different. The stringent limits on the number of cows and cattle, pigs, sheep and poultry which the collective farmer was allowed to keep were not nearly so restrictive in practice because of the poverty and diminutive scale of operations of the individual peasants before collectivization. In 1939 the average collective farm household contained as its personal possession not much less than 1 cow and 1 calf, 1 pig and $2\frac{1}{2}$ sheep or goats— absurdly little in all conscience, but in total by far the greater part of the country's shrunken herds and flocks:

TABLE 15

Private livestock in per cent of total—1934-41

January 1st	Cattle	Cows	Pigs	Sheep	Goats	Horses
1934	62	75	45	51	83	27
1935	62	74	58	51	83	20
1936	61	74	64	48	84	12
1937	60	74	57	47	84	10
1938	64	75	65	50	86	11
1939	64	76	62	53	86	9
1940	55	72	55	42	78	6
1941	57	75	58	42	75	19

Source: Calculated from Selkhoz, p. 264.

Under the impact of the mass slaughter of livestock during the mass collectivization of 1929-32, the Government looked at first to the private initiative of the peasants for the restoration of the national herd during the years of the Second Five Year Plan which contained hopelessly optimistic targets in this field. The relative security given to the collective farmers for their small livestock enterprises in the Model Charter of 1935 and the improvement in crops, particularly the bumper crop of 1937, gave a considerable impetus to the development of these enterprises.

The Communists reacted to this partial reassertion of the individualistic tendencies of the peasants by attempting to tighten the reins which they had loosened in an endeavour to improve the market supplies of agricultural products. They mounted an energetic campaign against the encroachment by the peasants' individual plots on collective farm land (Order of May 27, 1939).[39] At the same time they instructed each collective farm to set up at least two livestock farms, one for cattle, the other for pigs or sheep, and laid down a definite scale of the number of animals to be kept on collective farms of various sizes (Order of July 8, 1939).[40] The first of these measures made it more difficult for the peasants to feed their 'personal' livestock, the second compelled them to transfer some of it to the collective farms which had to stock their own livestock enterprises with the legally required minimum numbers. Hence the sharp reversal in 1939 of the previous trend towards a more than proportionate expansion in the personal livestock holdings. In that single year privately held stocks fell by 7.8m. cattle (including 1.7m. cows), 3.3m. pigs, 8.6m. sheep and 1.6m. goats, as a consequence of

114

the marked change in official policy towards the 'dual economy' of the collective farmers. (In 1940, the inclusion in the statistics of the newly acquired territories with their private agricultural system distorts comparisons.)

The official view of the dividing line between public and private agriculture was given by Stalin himself at the Second Congress of Collective Farmers, when he explained that it was 'better to proceed on the assumption that there is a collective farm economy, social, large and decisive, needed for the satisfaction of social needs, and there exists with it a small individual economy, needed for the satisfaction of the personal needs of the collective farmers'.[41] As so often with Stalin's pronouncements, this statement was seriously misleading as a statement of fact but contained important hints concerning the momentary policy of the régime. It was true enough that the State was taking the lion's share of the output of the collective farms in one form or another, but it was quite wrong to suggest that the 'subsidiary personal economy' of the collective farmers was simply part of a subsistence economy. In fact, the importance of the subsidiary economy for the market product of agriculture in certain fields was very great, and the tension produced by the virtual socialization of basic crops, including fodder supplies, and the largely private character of the major part of livestock production were among the main causes of friction and instability in the dual economy. By pointedly ignoring the part played by the private economy in supplying the market, Stalin indicated that the régime reserved the right to restrict the operations of the collective farmer in this field, if and when this suited the authorities.

This ambiguity in the official attitude towards the private economy as a source of market produce reflects the long tradition of friction and conflicting interests between the régime and the peasants as sellers of their produce. In the acute stage of the collectivization struggle the Government went so far as to prohibit direct sales by the peasants to consumers, but the attempt to channel the exchange between town and country exclusively through the public trade system had to be given up. In May 1932, at the depth of the agrarian crisis, the Government recognized the realities of the food situation by legalizing the free market within certain limits under the name of 'collective farm market'.

The new system was based on the direct exchange between producers and consumers; middlemen in the form of professional traders buying from the peasants and specializing in trade remained permanently forbidden. The supply was not limited to the surpluses of the subsidiary private economy. On the contrary, the name of the new institution indicated the part which the collective farms themselves were to play in it. The first claimant on their market produce, after payment in kind for the services of the MTS, remained the Government quota at almost nominal prices. In addition, they were encouraged—partly by official pressure and partly by the lure of preferential treatment —to make contracts at somewhat higher prices with the public trade system for supplying part of their production in exchange for scarce industrial goods at relatively favourable prices. Any remaining surplus they were free to sell in the collective farm market at free market prices and to use the proceeds to increase their cash income.

The importance of the collective farm market as a source of income for the collective farms can be seen from the fact that in 1940 they obtained 7,300m. rubles from free market sales compared with 9,100m. rubles from Government procurements.[42] As it will be seen below that Government procurements accounted for almost all the grain and technical crops, and the larger proportion of the potatoes, vegetables and animal products marketed by the collective farms, these figures give some indication of the enormous disproportion between free market prices and those which the collective farms (and the peasants) received from the public trade system.

In real terms it was calculated that the collective farm market accounted for 17.2 per cent of total urban retail sales in 1937 and for 12.5 per cent in 1940.[43] An approximate attempt to estimate the marketable surplus of the public and private sectors of agriculture for individual commodities in 1940, and to distinguish between Government procurements and free market sales, though valid only in broad terms, can be based on the official statistics.

1940 was not a typical year for many reasons, including the

116

TABLE 16

Agricultural gross production and market production—1940
(m. tons, eggs '000m. units)

Product	Gross production	Total Market	State procurements	Other market production Total	Public	Private
Grain	95.5	38.3	36.4	1.9	1.7	0.2
Raw cotton	2.24	2.24	2.24	—	—	—
Sugar beet	18.0	17.4	17.4	—	—	—
Sunflower seed	2.64	1.87	1.5	0.37	?	?
Potatoes	75.9	12.9	8.46	4.44	0.78	3.66
Vegetables	13.7	6.1	2.97	3.13	2.27	0.86
Meat and fat*	7.5	4.2	2.17	2.03	0.67	1.36
Milk	33.6	10.8	6.45	4.35	1.05	3.30
Eggs	12.2	4.7	2.7	2.0	0.14	1.86
Wool	0.16	0.12	0.12	—	—	—

*Live weight.
Source: Selkhoz, *passim* and calculations therefrom.

fact that agriculture in the newly appropriated territories was not yet fully integrated into the Soviet system and the proportion of 'private' market supplies for the collective farm market may be somewhat above average. But this should not unduly affect the main conclusion that the private sector was a major source of potatoes, vegetables and all animal products for the supply of the town population and was to retain its importance for a long time to come.

Apart from the overriding political considerations which made a compromise solution inevitable, there were reasonable economic grounds for the form of the dual economy in the later 1930s. The economic case for the large units of public agriculture was the advantage of mechanization for the more effective cultivation of staple crops; whether the climate in which collectivization was carried out made it possible to reap this advantage at the time was a different question, but one which is not necessarily decisive as a basis on which the policy as a whole should be judged. On the other hand, the technological level of animal husbandry was (and in many respects still remains until today) abysmally backward, based entirely on manual labour and without appreciable economies of scale. The economic case for leaving animal husbandry organized in relatively small units was, therefore, strong—but it did not extend to the policy

117

actually pursued, which dispersed its major part to such an extent that it became a mere back-yard enterprise.

On the marketing side, the co-existence of strictly controlled public trade and more or less free collective farm trade was not only a necessity in the conditions of the time but had certain positive advantages for the authorities. It provided the peasants and the collective farms with a fairly strong incentive to expand market production and thereby to improve both their incomes and the inadequate food supply of the towns. In addition, it acted as a sponge absorbing a large part of the inflationary pressure caused by the insufficient supply of all kinds of consumer goods in a period of rapidly rising money incomes.

The outstanding weakness of the collective farm market arrangement was the incredible waste of labour involved in the personal sale of tiny quantities of produce by the producers themselves. The political risks of permitting the rebirth of an uncontrollable layer of middlemen were obviously—and, on the whole, probably rightly—unacceptable to the Government, but it took many years before a constructive alternative was found in the institution of the so-called commission sales during the post-Stalin era.

Output and Income Trends—1933-40

One of the first casualties of the great crisis in the countryside was the volume and accuracy of statistical information. The foremost example of this process was the falsification of grain-production figures, and in view of the key position of grain supplies for the volume of agricultural production and the food supplies of the population this was a matter of great practical importance. It was, indeed, noted at the time that the decision to express harvest estimates in terms of 'biological' yield was designed to produce inflated estimates, but the extent of this bold-faced manoeuvre was estimated correctly by very few people outside Russia; more important, there is reason to believe that it may have deceived even the Soviet bureaucracy itself.

In his report to the 18th Congress of the Communist Party in March 1939, Stalin provided figures for the grain crops of the Second Five Year Plan. They were 89.9m. tons in 1933, 89.4m. tons in 1934, 90.1m. tons in 1935, 82.7m. tons in 1936 and

120.3m. tons in 1937.[44] The total for the five years was, therefore, 472.4m. tons and the average per year 94.5m. tons. Though not very impressive compared with the 1933 targets of the First Five Year Plan of 99.6-105.8m. tons, this would have been an increase of almost 30 per cent on actual gross production in 1927-28 and a tolerable basis for the official claim that the reorganization of agriculture on a modern basis had been accomplished.

However, when the statistical blackout was lifted during the Khrushchev era, the average barn yield for 1933-37 was quoted as 72.9m. tons (Appendix, Table II), marginally less than in the disastrous years of the First Five Year Plan and even a shade below the 1927-28 crop. These revised official crop estimates would indicate that the average grain crop during the Second Five Year Plan had been overstated by as much as 30 per cent. Contemporary claims by the Soviet authorities must, therefore, be regarded as baseless.

As for other crops, the position was distinctly patchy. The production of raw cotton and sugar-beet expanded considerably, that of potatoes and vegetables stagnated, while oil seeds and flax seem to have declined. (Appendix, Table II.) Such progress as there was had been achieved mainly by an expansion in the sown area, for unit yields in 1940, a very good year, were in some cases actually lower than in 1913, which had also been an exceptionally favourable year.

TABLE 17

Selected crop yields 1913 and 1940
(quintals per hectare)

	1913*	1940
All grains	8.2	8.6
Raw cotton	10.8	10.8
Sugar-beet	168	146
Flax fibre	3.2	1.7
Sunflower seed	7.6	7.4
Potatoes	76	99
Vegetables	84	91

* Contemporary borders.
Source: Selkhoz, p. 208.

The output of animal products fell to extremely low levels during the years following the collectivization crisis and slowly recovered afterwards, but remained generally below the far from impressive 1928 levels. (Appendix, Table III.)

Reliable information about the changes in living standards in town and village during the 1930s is, of course, impossible to come by, but the importance of the subject makes it necessary to review the scraps of evidence available. The compromise settlement with the peasants seems to have made it possible for them to consume a fairly high proportion of the livestock products which they produced in their subsidiary private economy and to obtain scarcity prices for much of their marketable surplus, after they had complied with their obligations to supply the Government with part of their production. Given the insatiable unsatisfied demand of the urban population for food, the peasants seem to have fared at least better than the workers in the towns.

A very crude estimate of the changes in urban population and market produce between 1913 and 1940 suggests that supplies of potatoes and vegetables per head of the urban population in 1940 were rather larger than a quarter century before and those of grain perhaps equal, if allowance is made for the absence of exports and the reduced number of horses in 1940; on the other hand, supplies of meat, milk and eggs were only between 50 and 75 per cent of the 1913 level. Such crude estimates ignore the effects of official stock-piling in 1940 which was, of course, a very abnormal year, but they provide a very rough measure of the deterioration in urban food supplies. On a similar basis, the supplies per head of the farm population in 1940 were about the same for grain and higher for potatoes and vegetables than in 1913, marginally higher for milk and eggs and perhaps a quarter lower for meat. Though subject to even stronger qualifications than the calculations for the urban population, such figures suggest that the peasants may have been rather more successful than the urban workers in protecting their material standard of living during the period immediately preceding the Second World War.

The effects of inflation during the 1930s on retail prices make the construction of an index of real wages extremely hazardous (and the choice of a base period crucial for the interpretation of the results), but the following estimates clearly indicate, even if they do not accurately measure, the fall in living standards of the non-agricultural population throughout this period:

TABLE 18

Index of annual real wages (1937=100)
(including social benefits)

	based on prices in	
	current year	1937
1928	116	165
1937	100	100
1940	90	94

Source: Janet G. Chapman: *Real Wages in Soviet Russia since 1928* (1963), p. 144.

Such calculations are even more difficult to make for the peasants, and their results are subject to even stronger reservations. In 1932, the value of a labour day averaged 74 kopeks, including 2 kg. of grain, and the average income of a collective farm household from the collective farm was 184.2 rubles.[45] According to Jasny's estimates, the fall in real income per head of the peasants between 1928 and 1937 was of the order of 19 per cent, compared with about 40 per cent for non-agricultural wage earners.[46] Real incomes declined further between 1937 and 1940, when those of the peasants were only two-thirds of the 1928 level and those of non-farm workers only a little more than one-half.[47] From a somewhat different angle, other investigators came to the conclusion that the earnings of collective farmers rose between 1937 and 1940 by 19 per cent relative to those of hired workers,[48] which is of course compatible with a general decline in absolute living standards with the approach towards the outbreak of war.

However problematical the individual calculations, the direction of the changes seems to be reasonably well established. During the later thirties the peasants were almost certainly doing relatively better (or at least less badly) than the industrial workers. More than ten years after the great debate about the nature of primitive socialist accumulation it might well have been concluded that the great war of the régime against the peasants had ended in stalemate, and that the proportionately biggest contribution to the forced industrialization of the country was being extracted from the workers rather than from the peasants.

An interesting inferential piece of evidence in support of this conclusion is the slowing down in the rate of population movements from the village to the towns. During the NEP period the towns had been swamped by the immigration of large numbers of ex-peasants who caused persistent unemployment despite steadily rising industrial production. Ten years later, Stalin claimed that 'unemployed peasants who have no roof over their heads, who have left the village and are threatened by hunger' had ceased to exist in the Soviet Union. 'Today it can only be a question of asking the collective farms to fulfil our request and to give us for our growing industries annually at least half a million young collective farmers.'[49] Even when the official gloss has been removed from this statement, it retains some significance as indicating a definite change in the relative position of the peasant and the urban worker on the eve of the Second World War, compared with that at the start of the reconstruction period.

3. THE BALANCE SHEET OF COLLECTIVIZATION

What was the net contribution of the second agrarian revolution to the process of industrialization for the sake of which the Communists had carried through this great upheaval, and what were the conditions of sustained agricultural progress at its end?

Collectivization and Industrial Growth
The intractable issues which confronted the Communists towards the end of the NEP period had been discussed *ad nauseam* and from every angle in the great debates of the 1920s on the economic future of the Soviet Union. The First Five Year Plan embodied the ultimate decision in favour of forced expansion of heavy industry which involved a large contribution from agriculture to the enormous long-term investment needed for the construction of factories and mines and the import of machinery.

A very substantial expansion of agricultural output, and particularly of market production, was therefore an indispensable condition of the whole process. The authors of the Plan were honest enough to state their dilemma, and that of Soviet society as a whole, in plain terms; they knew that agricultural production had been stagnating since 1926 and that there were some unmistakable signs of a movement in the wrong direction during 1927 and 1928. For this reason they asked themselves: 'Have we got in such conditions a basis for planning such rates of development of agriculture as we are aiming at in the Plan, is this not a break with reality, a jump into the realm of imagination?'[50] They attributed the current agricultural difficulties of the régime mainly to the growth of the *kulaks* and hoped that their ambitious aims would be realized through a combination of two factors: higher productivity through the establishment of agricultural co-operatives and the development of a native capital goods industry which would make it unnecessary to rely on imported agricultural machinery and would make it possible to mechanize agricultural production.

The disappointment of these hopes could not have been more cruel. The regressive output trend of the late 1920s was intensified by forced collectivization and by the incompetence of the régime in managing the newly-created large-scale agriculture. In practice, the contribution made by agriculture towards industrialization consisted essentially in the extortion of an excessive share of its output for feeding the towns and for export, without a corresponding increase in the supply of industrial goods to the peasants.

Despite the growth of industrial production, agriculture remained—and remains even today—the main source of the goods entering into mass living standards either in the form of foodstuffs or in that of simple consumer goods (textiles and shoes). The sharp fall in food production, and particularly in animal husbandry, was thus directly reflected in declining living standards. The shortage and poor quality of food was also one of the major causes of the lagging behind of the productivity of industrial labour, despite the heavy investment in mechanization and modernization of industrial plant. Ill-fed workers, compelled to supplement inadequate rations—obtained by hours of queuing—through devious transactions, were incapable of sustained efforts and moved from job to job at very little provocation in search of better conditions. At the same time, the insufficient supply of essential raw materials for the food industry, and to a lesser extent for other consumer goods industries, hampered their growth and was one of the main reasons for their chronic failure to reach the planned targets.

Perhaps even more damaging to Soviet society were the social consequences of the economic policy of the Soviet régime. A rigid stratification of political positions is, of course, implicit in a bureaucratic dictatorship as such, but the impoverishment of the country in terms of food and consumer goods added economic discrimination to political domination. With the mass of the people reduced to subsistence level, the groups whose efficiency, and loyalty to the régime, were particularly important had to be shielded as far as possible from the effects of the policy which was responsible for this state of affairs; this applied to skilled workers and managers in industry, the Communist Party hierarchy and, above all, the political police which became increasingly the main pillar of the Government and obtained valuable economic privileges to buttress its superior social status.

In the last resort, it was the agrarian policy of the régime and its total failure as a production strategy which appears as the biggest single factor in this debacle. It was at the root of many of the economic difficulties and of the social degeneration of the 1930s which culminated in the great purges.

The New Conditions of Agricultural Progress
In every respect but one the official attitude towards the peasants during the second agrarian revolution was self-defeating and irrational; its key element was the determination to bring about a radical change in the balance of power between the régime and the peasants. Perhaps the most favourable verdict about its other aspects is that it succeeded in the years after 1933 in repairing some of the damage caused during the period of compulsory mass collectivization.

The most important long-term consequence of this painful process was the partial substitution of large and medium productive units for the multi-millions of very small peasant holdings in the sphere of crop cultivation (except potatoes, fruit and

124

vegetables) which was more than ever the mainstay of the rural economy after the mass destruction of livestock between 1929 and 1932.

In the traditional small-scale agriculture of pre-collectivization days, land and capital had been very scarce and labour generally very plentiful. Management in the sense of a deliberate manipulation of the factors of production for the purpose of obtaining the highest margin over costs was generally limited by the determination of the individual peasant to do the best he could within the accepted farming system, which generally included adherence to a primitive crop rotation sequence. Increase in output was, therefore, mainly dependent on the more intensive use of labour, supported by the re-investment of parts of the products of labour in more tools and such simple machinery as could be afforded and employed on dwarf holdings. The most promising branch of farming for the more intensive use of labour was animal husbandry, and livestock was, therefore, in many parts of the country the visible embodiment of prosperity. The tendency was thus towards a mixed farming system with an increasing bias towards animal husbandry where climatic conditions did not make this impossible. The immediate result of mass collectivization in near-civil war conditions was a violent reversal of this trend.

Collectivization, with mechanization of ploughing and harvesting as its main technical instrument, had some effect on land usage, but its most important influence was on labour and management. The amalgamation of the fragmented plots of individual small-scale producers into substantial fields which made mechanical cultivation possible and, indeed, highly desirable increased the effective land supply and should have resulted, other things being equal, in higher yields. Other things were far from equal; ignoring the fundamental change in the distribution of agricultural net income between the peasants and the State to the detriment of the peasants, the most critical difficulty was the absence of competent management on the required scale.

The task of finding within a few years, or even months, the skilled personnel needed to run more than two hundred thousand collective farms and some thousands of unwieldy State farms was, of course, insoluble and the result was incompetent management on an enormous scale. The despatch of reliable Communists from the towns to the countryside as chairmen of newly established collective farms consisting of bitterly resentful and hostile peasants was an act of political, or even of military, warfare, and the enthusiasm of these men was no compensation for their managerial weakness.

The bureaucratic defects of the system of government and administration compounded the difficulties of the local representatives of the régime. The tight control of the Ministry of Agriculture, of the local Government officials and of the Party bosses compelled farm managers to comply with instructions which were often wrong and always far too detailed for successful application in a field where intelligent response to changing conditions is vital to good management.

Mechanization considerably reduced the need for manual labour at the critical periods of the agricultural year, sowing time and harvesting time, but labour shortage had not been the most pressing problem in the old days. (On the other hand, mechanization did not go nearly far enough to concentrate the harvesting operation into the often very short period available in Russian climatic conditions for a really satisfactory harvest, and plant breakdowns further reduced the effective value of the available machinery, thus causing severe reductions in yields.) At the same time, mechanization needed much more skilled labour than had been needed before and such labour was (and remains) in short supply. The Machine and Tractor Stations had a staff of more than half a million people, amongst them many with very scarce skills, such as engineers and mechanics. Whether the net result was a better deployment of the available labour resources cannot be decided in general terms, because it depended on the use made of the redundant farm labour. During the collectivization crisis much of it was used cruelly and wastefully in labour camps run by the political police which remained a prominent feature of the Stalin era as a whole. Other ex-peasants supplied a large proportion of the labour needed for the enormous construction projects in industry and transport and manned the manufacturing plants—almost certainly to the benefit of Soviet society and probably to that of the individuals concerned. Finally, for the bulk of the collective farmers mechanization of land cultivation in the collective farms increased the opportunity of tending their individual plots and livestock.

From a technical point of view one of the most obvious weaknesses of the new large enterprises was their lack of specialization. Despite the theoretical recognition of the advantages of specialization, the collective farms in particular tended to reproduce the farming system of the near-subsistence holdings which they replaced, and this trend became even more pronounced towards the end of the pre-war period, when the compromise settlement of 1935 came under considerable strain and collective farms and State farms were instructed to raise more livestock in order to reduce the dependence of the régime on the privately produced livestock products of the individual plots: '. . . during the period of the personality cult (i.e, in the Stalin era) . . . a " theory" of uniformity of the geographical distribution of agriculture by zones and regions of the country was advocated. Emphasis was laid on such a distribution and specialization as would enable each part of the country to produce all the essential agricultural products it needed. Large-scale propaganda was also made for the principle of the diversification of collective and State farms, with compulsory setting up of four or more livestock units (*fermy*), independent of actual production conditions.'[51]

It was easier to criticize such a policy than to follow the path of specialization in the absence of the necessary conditions for its success. Specialized units have to be able to rely implicitly on the regularity and adequacy of outside supplies, industrial and agricultural materials as well as finished products. The price for the economies of scale and the rise in productivity obtainable in suitable conditions is a great increase in the dependence of the individual enterprise on the rest of the economy. A specialized farming system therefore needs a highly developed distribution network, both at wholesale and at retail level, and adequate stocks at every stage of distribution. In this it is basically similar to modern industry, but the physical conditions are different: distances are longer, the economic density of operations is lower and the number of links in the chain tends to be larger. Hence there is a greater risk of breakdown which can be effectively countered only if substantial stocks are maintained at a large number of points.

Last but not least, specialized large-scale agriculture depends at every step on industrial supplies. At the near-subsistence level of NEP agriculture, the peasant supplied the labour and his livestock (horse or cattle) the draft power. Both lived on the produce of the land, though the peasant supplemented his diet by the fruits of his animal husbandry and obtained a modicum of supplies from industry. In the large-scale public farm enterprise, much of the labour and most of the energy is supplied by industry. Skilled mechanics and engineers as well as machine operators supplement peasant labour, tractors, drills and combines replace most of the animal draft power and much human labour. Instead of feeding on the produce of the land they till, they have an insatiable thirst for oil and cannot carry on for any length of time without tyres, belts and a whole host of spare parts. If in exchange for their services they also take part of the harvest, this is not needed for direct consumption but in payment for the industrial products they use and for the investment made by the State in providing them. Nor are their operators satisfied with peasant standards of living but require much higher rates of pay and consumer goods on which to spend it.

In the years immediately following the second agrarian revolution the economic and technical conditions for thoroughgoing agricultural specialization were not in existence, but the partial integration of the village into the complex processes of a modern economy made agriculture much more dependent than in the past on industry. The consequences of this situation were made dramatically apparent by the war in which the Soviet Union became involved through the German attack in June 1941.

The stabilization of the collective agriculture emerging from the second agrarian revolution had made sufficient progress at the end of the pre-war period to justify the belief that it was viable in principle but woefully inefficient in practice. If it was to overcome its still serious weaknesses, it would need large resources of modern equipment and current industrial inputs, together with skilled labour for its operation; competent management and a radical overhaul in the system of central control over its activities; and a stable balance between collective and private interests which presupposed a change in the distribution of the net income of the collective farms in favour of the peasants, in order to make it worth their while to put as much energy and initiative into their work for the collective economy as in the exploitation of their subsidiary private plots.

VII
War, Recovery and Stalemate (1941-1953)

1. THE SECOND WORLD WAR

The simpler and less differentiated an economic system, the greater its resistance to external pressure and catastrophe and its powers of recuperation. The recovery of the post-revolutionary peasant economy from the crushing burdens of the First World War and the Civil War was surprisingly quick : there was famine in 1921 and production at or near the average pre-war level four years later. When war broke out again in 1941, the old peasant farming had been replaced by a complex large-scale agriculture. Although the State farms were a significant factor in the new system, the main element in the situation was the co-existence of the collective farms and of the private subsidiary economy of the peasants on terms which embodied a tolerable but precarious compromise between the interests of the régime, the urban consumers and the rural producers.

The physical emergency of the Second World War lasted from beginning to end less than four years (compared with well over six for the First World War and its even more destructive sequel), and for most of the occupied territories only two or three years. Large areas were never directly exposed to military action in the way in which most of Russia had been affected either during the previous war or in the period of civil war and intervention, but the return to peace conditions and economic advance proved to be much more lengthy and arduous than during the 1920s.

The direct consequences of the Second World War were certainly very grim and the bare outline of the decline in human and material resources, though only incompletely known, is depressing in the extreme. The official Soviet claim, made by Khrushchev,[1] of a loss of twenty million lives as a result of the war is only too credible, although it cannot be much more than a broad guess. Between 1939 and the beginning of 1950 the population on a somewhat larger territory (because of post-war acquisitions) declined by 12.2m. people; the fall must have been very much greater between 1941 and 1945. On balance, this fall was borne entirely by the rural population which stood at 130.3m. in 1939 and at 109.1m. in 1950, a drop of 21.2m., while the town population rose during the same period by 9m.[2] The bulk of the decline must have taken place on the farms which are generally assumed to account for about 85 per cent of the rural population. Migration from the countryside to the towns is, of course, a normal feature of economic development, but conditions during this period were far from normal.

Such global figures do not give an adequate indication of the quality of the population losses incurred by the Soviet village as a result of the war, when military service must have fallen most heavily on the peasantry, the traditional man-power reservoir for the lavish strategy of the Soviet high command as for that of its Tsarist predecessors. Soviet agriculture at the end of the war was disproportionately weakened by the huge loss in young men; the women thus had to endure, in addition to their emotional misery, a correspondingly heavier economic burden. By 1942, 43 per cent of all combine operators and 46 per cent of all tractor drivers in the MTS were women, although their share of these occupations had been insignificant before the war.[3]

In one Ukranian collective farm for which detailed figures are available, no fewer than 125 households out of 336 were entirely without men in 1947. The total population of 1,358 included 335 children under ten and 1,023 persons over ten years of age of whom there were 68 boys and 91 girls between ten and sixteen as well as 250 men and 614 women. The farm labour force had only 71 able-bodied men under sixty and 284 able-bodied women under fifty.[4] This farm was in territory occupied by the Germans for much of the war and its losses in man-power may, therefore, have been above average, but the situation was stark indeed.

131

Production

The area occupied by the Germans in November 1941, before some of the deepest enemy penetration of the Ukraine, produced before the war 38 per cent of the gross grain harvest, as well as 84 per cent of total sugar-beet production and contained 38 per cent of the national cattle herd and 60 per cent of the total number of pigs.[5] Altogether the occupied territories accounted for 41 per cent of the sown area of the Soviet Union and over 45 per cent of total livestock units.

The official output indices, for what they are worth, show a fall of 40 per cent in gross production, spread almost evenly amongst crops and livestock products between 1940 and 1945. (Appendix, Table I.) The area under grain fell from 110.4m. hectares in 1940 to 67.3m. in 1942; by 1945 it had recovered to 85m.,[6] a loss of about a quarter, while the decline in yields may have been of a similar order of magnitude. On the reduced area it was attempted as far as possible to compensate for the temporary loss of some of the best land in the Ukraine and the Kuban by growing more wheat in the unoccupied territories east of the Volga and in Siberia.[7]

The acreage figures for fodder crops and technical crops show similar, and partly even heavier, falls, and only the potato and vegetable acreage was claimed to have been restored to pre-war levels by the end of the war.

TABLE 19
Sown areas—1940–45
(million ha.)

	1940	1942	1944	1945
Cereals	110.4	67.3	82.0	85.0
Technical crops	11.8	5.9	7.5	7.7
Fodder crops	18.1	9.5	10.4	10.1
Potatoes, vegetables	10.0	5.0	10.0	10.0
Total	150.4	87.7	109.9	112.8

Source: D. I. Shigalin: *Narodnoie Khoziaistvo v period velikoi otechestvennoi voiny* (Moscow 1960) p. 197.

The good showing of potatoes and vegetables was due to two factors: they were grown extensively on private allotments, and

132

Siberia, the Urals, the Far East, Central Asia and particularly Khazakstan substantially increased the areas devoted to these products.[8]

Between 1940 and 1945 grain production fell by just over one-half, from 95.5m. tons to 47.3m., sugar-beet production declined from 18m. tons to 5.5m. and that of raw cotton from 2.2m. tons to 1.2m. Despite the recovery in acreage, potato production in 1945 was only 58.3m. tons compared with 75.9m. before the war and that of vegetables 10.3m. tons compared with 13.7m. (Appendix, Table II.) Supplies which had been barely adequate before the war were thus totally insufficient at its end, though in the case of potatoes—regarded as 'second bread'—direct consumption as human food increased and the shortfall was fully felt only in stock feeding.

The livestock figures tell the same dismal story. (Appendix, Table IV.) If the statistics are to be believed, the most striking feature of the cattle figures is their moderately good showing. The sharpest decline was naturally in pig numbers and perhaps just as characteristic is the almost unchanged number of goats whose toughness and ability to survive in the least favourable conditions is paralleled by their doubtful economic value. The halving of the already sharply reduced number of horses reflects partly their importance for Soviet military transport throughout the Second World War and partly the grave draft-power problem facing Soviet agriculture at its end.

Global figures inevitably obscure the differences between areas and throw no light on the special plight of the ex-occupied territories. In the early stages of the war, much of the collectively-owned livestock in danger areas was driven east, while most of the remainder was killed rather than allowed to fall into German hands. Nevertheless it was claimed that the enemy killed or removed huge numbers of livestock in the areas temporarily under their rule—no fewer than 7m. horses (out of a total of 11.6m. in the afflicted territories), 17m. cattle (out of 31m.), 20m. pigs (out of 23.6m.) and 27m. sheep and goats (out of 43m.).[9] In the districts not directly affected by hostilities the numbers of cattle, sheep and goats—but not of pigs—remained very close to the pre-war level.[10] The huge drop during the first two war years and the subsequent recovery suggest that the emergency in this field was due more to the direct action of the Germans than to the economic consequences of the war. Whatever the reason, in the event the fall in livestock herds and flocks not only endangered the food supply of the population but greatly weakened the productive system of agriculture.

Equipment

The reduction in the animal draft power provided by horses and oxen was compounded by the deterioration of mechanical traction. Not only were tractor factories switched to tank production, but the two great plants at Kharkov and Stalingrad were out of commission for long periods. Direct enemy looting played its part in this field, too; it is claimed to have been responsible for the loss of 137,000 tractors, 49,000 combine harvesters, 46,000 drill ploughs and 35,000 threshing machines.[9] In terms of total horse-power, the energy resources of Soviet agriculture in animals and machines fell from 47.5m. before the war to 28m. at the beginning of 1946.[11]

At the same time the available machinery became less and less effective through obsolescence, lack of spare parts and shortage of fuel; in 1944 the average performance per 15-h.p. tractor was only 241 ha. of standard ploughing compared with 411 ha. before the war[12]—a decline of almost 40 per cent. Although detailed plans for the return of livestock and machinery to the liberated areas were made as early as the summer of 1943, the situation remained worst in these territories where disorganization was rife and losses of draft animals had been heaviest, quite apart from the loss of 2,890 MTS put out of action by the Germans.[9]

The consequences of the deterioration in the functioning of the MTS in these territories were particularly serious, since the collective farms had to rely on them more completely than elsewhere because of the depletion in their own resources: 'All the equipment was in bad condition, and the more complex the equipment, the worse its condition. There was a lack of spare parts and of shelter to protect the equipment from the weather. Many machines had been damaged by being used for tasks for which they were unsuited. . . . The shortage of spare parts was so great that during the main agricultural campaigns all valuable equipment, particularly tractors, had to be guarded whenever not in use; otherwise members of other farms or other MTS would steal vitally needed parts; for instance, on one occasion in 1948, all the wheels were stolen from our cultivators in the middle of the night, and nobody knew who took them.'[13]

Trade, Incomes and Prices

Wars are traditionally favourable for primary producers, and the Russian peasants seem to have obtained some relative advantage from their position during the Second World War. But though some concessions to them were inevitable, the official machinery of control and exploitation survived the test of wartime confusion and stress. The immediate consequence of the German attack was, indeed, a tightening of discipline, an increase in the number of obligatory labour days for each collective farm member and in deliveries of meat and wool, as well as the introduction of an additional delivery quota for meat and grain for the Red Army.[14]

In global terms, agricultural gross production fell by 40 per cent between 1940 and 1945 and market production by 43 per cent, with a higher proportion of vegetable products supplied by agriculture to the rest of society and a fall in the proportion of animal products: the gross production of vegetable products fell by 43 per cent and their market production by 38 per cent, while for animal products the fall in gross production was 'only' 36 per cent and in market production 53 per cent.[15]

The pressure for the highest possible deliveries was thus most effective for grains and, to a lesser extent, for potatoes; the proportion of sugar-beet marketed actually fell quite substantially, as did that of all the more important livestock products.

In September 1946 the authorities issued a special order for the purpose of retrieving collective farm lands appropriated by the peasants for their private use. This suggests a certain shift in favour of the peasants' personal economy during the war, though this does not seem to have gone to any great lengths. The most spectacular but least substantial advantage accruing to the peasants during the war and its aftermath was, however, the fabulous rise in the prices of their products in the collective farm market; in practice this was a problematical benefit,

TABLE 20
The fall in market production during the Second World War

	Total market production*		Market production in % of gross production	
	1940	1945	1940	1945
Cereals	38.3	23.2	40	49
Sugar-beet	17.4	4.7	96	86
Cotton	2.24	1.16	100	100
Sunflower seed	1.87	0.59	71	70
Potatoes	12.9	11.8	17	20
Meat and fat†	2.6	1.3	67	52
Milk‡	10.8	5.4	33	21
Eggs	4.7	1.5	41	31
Wool	120.0	73.0	72	66

* All commodities are in million tons, except eggs ('000m. units) and wool ('000 tons).   † Dead weight.
‡ Milk and milk equivalent of milk products.
Source: Selkhoz, pp. 86ff.

because the supply of industrial goods to the village was so small that the peasants had to accumulate large amounts of paper money which was in due course deprived even officially of nine-tenths of its value by the currency reform of 1947.

In the conditions of the time the use of the printing press as a source of war finance was a necessary evil, and the most competent Soviet leaders in economic matters were under no illusion about its drawbacks: 'The amount of currency in circulation cannot increase to an unlimited extent. An excessive increase of currency in circulation can undermine the material incentives for growth in labour productivity, particularly in agriculture. Surplus currency in circulation creates an inflated demand and, not encountering a sufficient supply, lowers material incentives for raising the productivity of labour, and leads to an excessive price rise in the *kolkhoz* peasant market.'[16]

Price trends on the collective farm market were, indeed, a fever chart of the economy and reflected the growing disparity between money incomes and the supply of consumer goods, particularly food. In 1944 free market prices were about eleven times their 1940 level.[17] During these four years, turnover of public (State and co-operative) trade declined at current prices from 175.1 billion rubles to 119.3 billions or by almost a third, while collective farm market turnover increased from 29.1 to

33 billion rubles or by more than one-eighth, although the quantity of goods sold can hardly have been more than 10 per cent of pre-war.[18]

The turnover of public trade can be used only to a very limited extent as a measure of supplies reaching the urban and rural consumers; such as they are, the figures indicate that sales to the peasants were cut more severely than those to the urban population.

TABLE 21

Retail turnover of public trade, 1940 and 1945
(billion rubles at current prices)

	1940			1945		
	Total	Urban	Rural	Total	Urban	Rural
Goods	152.2	104.9	47.3	127.0	89.5	37.5
Public catering	22.8	18.5	4.3	33.1	28.9	4.2
Grand total	175.0	123.4	51.6	160.1	118.4	41.7

Source: Narkhoz 1962, pp. 515f.

Goods turnover fell by 15 per cent in urban trade and by 21 per cent in rural trade. Some of the worst deficiencies in the supply of foodstuffs to urban consumers were made good by the expansion in public catering which increased, in money terms, by 56 per cent in the towns but declined marginally in the countryside. As even in public trade prices in 1945 were higher than before the war (partly because of the opening of so-called commercial shops in 1944), the figures understate the contraction in volume. Even so, they suggest a deterioration in supplies to the village not only in absolute terms but also compared with the trend of supplies to the towns.

Physical output data of the consumer goods industries show more clearly the steep decline in the total quantities available, though they flatter the consumption of the civilian population by ignoring the priority given to the armed forces during these years.[19] The output of cotton fabrics in 1945 was only 1,617m. metres compared with 3,954m. in 1940. Socks and stockings fell from an inadequate 485m. pairs in 1940 to the pathetic total of 91m. pairs in 1945, underwear dropped from 124m. to 26.6m. and leather footwear from 211m. pairs to 63.1m.[20] The output of the food industries did not drop quite so steeply but nevertheless very substantially:

TABLE 22

Production of selected foodstuffs 1940 and 1945

	1940	1945	% decline 1945 over 1940
	(m. tons)		
Flour	28.8	14.6	49
Bread	24.0	11.0	54
Semolina	1.6	1.0	37
Sugar	2.1	0.5	78
Animal fats	0.23	0.12	50
Vegetable fats	0.8	0.3	63
Fish	1.4	1.1	20
Meat	1.5	0.7	56

Source: Shigalin, *op. cit.*, p. 181.

From whatever angle they are looked at, the facts show that by the end of the war the Soviet village had been bled white of essential human and material resources. In this respect it was by and large in a similar position to the country as a whole which emerged from the ordeal of war victorious beyond the wildest dreams of its leaders, but gravely debilitated. Thus the most critical question for the future was not whether the Soviet peasants had borne more than their fair share of sufferings and deprivations, but what their part in the economic reconstruction of the country was going to be and whether they were in a fit state to carry it out.

2. PRODUCTION AND MARKET SUPPLIES AFTER THE WAR

The Slow Climb-back

With the great power struggle between the régime and the peasants decided once for all and the invaders expelled from the country, the task before Soviet agriculture was the painstaking reconstruction of the gravely weakened system and the re-expansion of production to the pre-war level as a basis for the further growth needed to meet the greatly increased claims of the population and industry.

The recovery of output was slow and patchy. With 1940 as the last pre-war year as a base, gross production of agricultural products in 1953, the year of Stalin's death, and on a somewhat larger area, was only 4 per cent higher. This inadequate increase was entirely due to a rise of 24 per cent in animal husbandry, while crops were actually 4 per cent lower than in 1940, which had been a climatically more favourable year. (Appendix, Tables I, II and III.)

Although the accuracy of the official agricultural statistics has been questioned with good reason, the figures are broadly consistent. By 1940, most types of crops showed a net increase not only over the very depressed level of the years of the collectivization crisis but also over their pre-1914 levels, and the increase was impressive for most 'technical crops' (such as cotton, sugar-beet, oil-seeds, etc.) as well as potatoes, while livestock products had more or less recovered from the severe decline of the early 1930s. Between 1940 and 1953 the expansion in cotton and sugar-beet continued, animal husbandry made unspectacular but useful progress, while the grain crop and the production of potatoes and vegetables failed to increase. In fact, grain production reached its highest level during this period as early as 1949 and fell back in 1950; during the next three years it fluctuated within narrow limits and with no tendency to rise.

Given the fundamenal importance of grain for the food and livestock economy of the country, this obviously alarming state of affairs needs further examination.

TABLE 23

Gross crop of the main cereals (1913–53)

	1913	1940	1950	1953	m. tons	%
		(m. tons)			Change 1953 over 1940	
Wheat	26.3	31.7	31.1	41.3	+ 9.6	+30
Rye	22.7	21.0	18.0	14.5	− 6.5	−31
Other food grains	4.0	6.0	3.2	3.7	− 2.3	−38
Total food grains	53.0	58.7	52.3	59.5	+ 0.8	+ 1
Oats	17.0	16.8	13.0	10.0	− 6.8	−40
Barley	12.1	12.0	6.4	7.9	− 4.1	−34
Maize	2.1	5.1	6.6	3.7	− 1.4	−27
Other feed grains	1.8	2.9	2.9	1.4	− 1.5	−52
Total feed grains	33.0	36.8	28.9	23.0	−13.8	−38
Total cereals	86.0	95.5	81.2	82.4	−13.0	−14

Source: Selkhoz, pp. 202f.

Amongst food grains there had been a radical shift from rye, millet and buck-wheat to wheat, particularly since 1950. Gross production of food grains as a whole in 1953 was, however, only 1 per cent up on 1940 and this was entirely due to an expansion of 5½ per cent in the sown area. The yield of wheat was somewhat higher than before the Second World War, but this was more than balanced by falls of 20 per cent or more in the yields of rye and the minor food grains. As the population in 1953 had only just regained the pre-war level, gross production of food grains per head was very similar to what it had been in 1940 and there was no need for large emergency stock-piling such as had taken place in the immediate pre-war period.

The full force of the fall in grain output was felt in the main fodder crops which produced almost 40 per cent less than in 1940, to some extent due to the smaller acreage, particularly of oats, but largely as the result of savage falls in yields which were not only lower than in 1940 but lower than in 1913. As climatic conditions made the use of concentrated feeding stuffs essential for most of the year, the lower supply of feed grains was bound in due course to limit the growth of the livestock industry, although the official emphasis on the importance of grassland tried to counteract this effect.

The grassland rotation policy connected with the name of V. R. Vilyams (Williams)[21] was first advocated by its author before the Revolution and accepted by the leadership as early as the 1930s, but it reached its greatest practical importance after the Second World War. Its central thesis was the claim that crops depended, above all, on the soil structure and that natural grass played a crucial part in its improvement and the maintenance of soil fertility.

The system of grassland management, which was connected with the introduction of a complex eight- or ten-year rotation system, was particularly acceptable to the authorities because of its low claims on capital investment;[22] thus it seemed appropriate for the situation created by the neglect of agriculture and appealed to the régime much more than costly projects of intensification of agricultural production. It implied, however, that while capital was scarce the additional land needed for its successful application was plentiful and could be had without

substantial capital investment. The stagnation of grain production in the early 1950s indicated that this was not, in fact, the
case; at best its results would need another commodity which
was by then no less scarce than capital and additional fertile
land—time.

With the restoration of animal husbandry to its modest pre-
war level at the beginning of the 1950s, the main question was
how to feed the increasing herds and flocks, and the additional
grass available from the new cropping policy (either as hay or as
silage) was quite insufficient for this purpose. By 1953, the total
cattle herd exceeded the pre-war level by over a million, though
there were still 9 per cent fewer cows than in 1940. Pig numbers
were up by almost six million or more than one-fifth, and there
were an extra twenty-four million sheep and goats, up by more
than a quarter, and an additional hundred and fifty million
birds, an increase of over one-half.[23] Only the number of horses
continued its secular decline, and in 1953 it was almost six
million, or more than a quarter, lower than in 1940. (Appendix,
Table III.)

Fewer horses needed less oats, and the disappointingly slow
rise in the number of cows, which was castigated by Khrushchev
in his exposure of the agricultural situation in September 1953,
kept the demand for concentrates lower than it would have
been—though it could have been partly a response to the shortage of feeding stuffs. Sheep and goats did not make any special
claims on feed grains, but the growth in pig numbers was a
different matter, particularly in view of the stagnation of the
potato crop, and the growing flocks of chickens, geese, etc. also
needed growing supplies of concentrates to produce more eggs
and poultry meat.

Agricultural Output and Resources of the Main Sectors
The agricultural system of the Soviet Union consisted in 1940,
the last pre-war year, and in 1953, the year of Stalin's death, of
very similar elements—the public sector with its State farms,
collective farms and MTS, and the private sector which in 1940
comprised over a million independent peasants in the newly
acquired territories as well as the subsidiary personal allotments
of collective farmers and workers in the State farms and else-

141

where, while in 1953 the independent peasants had become integrated into the collective farm sector.

The contributions of these sectors to agricultural production can be expressed in reasonable detail in terms of gross production of the main commodities.

TABLE 24

Agricultural gross production by sectors, 1940 and 1953

(in m. tons, eggs in '000 m. units)

| | 1940 | | | | 1953 | | | |
| | | State* | Coll. | | | State* | Coll. | |
	Total	farms	farms	Private	Total	farms	farms	Private
Grain	95.5	7.5	76.2	11.8	82.5	9.7	70.5	2.3
Wheat	31.7	2.9	27.2	1.6	41.3	4.6	36.5	0.2
Other	63.8	4.6	49.0	10.2	41.2	5.1	34.0	2.1
Potatoes	75.9	1.6	25.2	49.1	72.6	2.6	17.9	52.1
Vegetables	13.7	1.2	6.0	6.5	11.4	1.6	4.3	5.5
Raw cotton	2.2	0.1	2.1	—	3.9	0.2	3.7	—
Sugar-beet	18.0	0.7	16.2	1.1	23.2	0.9	22.2	0.1
Sunflower seed	2.6	—	2.3	0.3	2.6	0.1	2.4	0.1
Hay	67.8	7.9	55.0	4.9	67.9	11.9	55.7	0.3
Silage	5.2	1.3	3.7	0.2	16.4	3.4	13.0	—
Meat and fat†	4.7	0.4	0.9	3.4	5.8	0.7	2.1	3.0
Milk	33.6	2.0	5.6	26.0	36.5	2.9	9.2	24.4
Eggs	12.2	0.2	0.5	11.5	16.1	0.5	2.0	13.6
Wool	0.16	0.02	0.08	0.06	0.24	0.03	0.17	0.04

* And other State enterprises. † Deadweight.

Source: Selkhoz, pp. 202ff, 334ff.

The point-to-point comparison of two such different years as 1940 and 1953 is subject to severe limitations, but produces some interesting results. The State farms showed an absolute increase in the gross production of every commodity in the table and in most cases also an increase in their share of total output, though this remained generally quite modest even in 1953. This performance supports the official estimate of an increase in gross production by about 40 per cent between the two years.[24] The collective farms, on the other hand, were in both years by far the largest producers of crops with the exception of potatoes and vegetables, but their gross production of these two important types of foodstuffs as well as of cereals, except wheat, was very appreciably lower in 1953 than it had been in 1940. On the other hand, their gross production of animal products was appreciably higher in 1953 than before the war. Their gross production as estimated by the official statistics had risen—but only by 7 per cent.[25]

The private sector had lost the limited importance in crop cultivation which it still retained in 1940, except for potatoes and vegetables, but despite the collectivization of the remaining independent peasants by 1953, the subsidiary plots of collective farmers and workers, etc., produced almost as much milk and meat and considerably more eggs than had been produced in 1940 by the independent peasants and the subsidiary personal economy together. Though its total production had fallen, no official estimate of its size is available.[56]

It is, however, possible to get a fair picture of the use of some important physical resources by the individual sectors by calculating their share of the area sown to some crops and of the national herd and comparing it with their share in the production of the corresponding products. Though a blunt instrument of analysis, such a comparison suggests some interesting conclusions (see Table 25).

In 1940, the State farms had the same share in the area sown to cereals and in the gross production of cereals; by 1953 both had increased, but the share of the State farms in gross production was now much higher than their proportion of the sown area. For potatoes and vegetables, the State farms produced in both years less than the proportion corresponding to their sown area, but the disproportion was much less in 1953 than it had been in 1940.

The collective farms also increased their share of the land devoted to cereals, potatoes and vegetables and their share of every kind of productive livestock and, with one exception, in the gross production of the commodities reviewed in the table above, but in almost all cases where such comparisons are possible the increase in their share of output was considerably lower than that in their land and livestock resources. For cereals, their proportion of the sown area rose between 1940 and 1953 by 7.2 per cent of the total but their proportion in cereal production increased only by 3.8 per cent of the total. (For potatoes and vegetables their share in output actually declined despite a rise of 6.5 per cent of the total in the area

TABLE 25

The distribution of land, livestock and production by sectors
(in % of the respective totals)

	1940 State* farms	1940 Coll. farms	1940 Private	1953 State* farms	1953 Coll. farms	1953 Private
1. Sown area						
Cereals	7.8	82.3	9.9	8.7	98.5	1.8
Potatoes and vegetables	4.6	42.8	52.6	6.1	49.0	44.9
2. Crops						
Cereals	7.8	79.8	12.4	11.8	85.4	2.8
Potatoes and vegetables	3.1	34.8	62.1	5.0	26.5	68.5
3. Livestock						
Total cattle	6.5	36.8	56.7	8.4	49.8	41.8
Cows	4.7	20.4	74.9	6.1	34.6	59.3
Pigs	11.9	29.9	58.2	13.9	40.8	45.3
Sheep	8.9	49.0	42.1	11.5	73.9	14.6
Goats	1.1	24.0	74.9	1.2	26.3	72.5
Poultry	0.7	10.5	88.8	1.9	17.7	80.4
4. Animal Products						
Meat—pigs	11.7	13.0	75.3	18.0	20.4	61.6
poultry	1.8	6.9	91.3	4.4	15.4	80.2
other	7.9	25.1	67.0	9.8	49.6	40.6
Milk	5.8	16.6	77.6	8.1	25.1	66.8
Eggs	1.6	4.1	94.3	3.1	12.4	84.5
Wool	11.8	49.1	39.1	14.0	70.7	15.3

* State farms include other Government farms except for poultry
numbers which refer to State farms only.
Source: Selkhoz 130f, 202ff, 270f, 322, 334ff.

sown to these crops.) Similarly, the share of the collective farms
in the total cow herd went up by 14.2 per cent of the total, but
their share of milk production increased only by 8.5 per
cent; their share of pig numbers rose by 10.9 per cent of
the total and their share in pig meat production by 7.4 per cent,
their share of the sheep flock went up by 24.9 per cent of the
total and their share of wool production by 21.6 per cent. Only
in the case of eggs, where their share went up by 8.3 per cent of
the total was the increase larger than that in the corresponding
share of the flock of animals, which rose only by 7.2 per cent of
the total, and the same applied, *mutatis mutandis,* to poultry
meat. As the output of comparable farm produce by the collec-

tive farms lagged behind their land and livestock resources even in 1940, the discrepancy thus widened quite appreciably with the course of the years.

The corollary to this relatively poor showing of the collective farms was not only the more favourable development on the State farms but also, and particularly, the further relative rise in the ratio between the land and livestock resources of the private peasant economy and their gross production of agricultural products. This took place simultaneously with the fairly general decline in their endowment with land and livestock resources and in the output of all products except potatoes and vegetables as a proportion of the total and could, indeed, have been connected with it.

These relationships are, of course, a different way of expressing differences in crop yields per acre and the yields of animal products per animal between the various types of agricultural enterprises in the Soviet Union.

TABLE 26

Yields of selected crops by sectors 1940 and 1953
(quintals per hectare)

| | 1940 | | | | 1953 | | | |
	Total	State farms	Coll. farms	Private	Total	State farms	Coll. farms	Private
All grain	8.6	8.7	8.4	10.8	7.8	10.4	7.4	12.7
Wheat	7.9	8.4	7.8	9.1	8.6	10.8	8.3	*
Rye	9.1	8.7	8.8	10.8	7.2	9.3	7.0	11.1
Oats	8.3	8.1	8.1	10.1	6.6	10.0	6.0	*
Potatoes	99	68	85	109	87	66	50	121
Vegetables	91	82	77	112	87	102	50	184

* Too insignificant production level to warrant yield calculations.
Source: Selkhoz, pp. 132, 136, 140, 202ff and calculations therefrom.

The important development was not the growth in the difference in yields between the public and the private sector. In the first place it is not really reasonable to compare garden cultivation and field cultivation, secondly a higher return per acre for small-scale cultivation is not peculiar to the Soviet Union but can be found in many countries and its evaluation depends to a large extent on the amount of labour spent on obtaining higher yields, which is generally high and often excessive. The really important change concerns, however, the relations between the two forms of large-scale public agriculture, the State farms and the collective farms. In a large country such as the Soviet Union average figures reflect to some extent differences in regional structures and must, therefore, be compared with great caution. Nevertheless it is clear that in 1940 yield levels, at least for the commodities shown in Table 25, were broadly similar in State farms and collective farms, with the balance of advantage fluctuating from product to product. (The higher wheat yield of the State farms was, in fact, limited to winter wheat; for spring wheat State and collective farms had the same very low yield of 6.6 quintals per ha.) By 1953, yields in the State farms had increased for all commodities shown above except potatoes, where they were marginally lower than in 1940. The collective farms, on the other hand, showed a modest improvement for wheat but startling falls for other grains, potatoes and vegetables, which were large enough in view of the preponderance of collective farms in Soviet agriculture as a whole to depress average yields for all these commodities except wheat. This makes it possible to conclude that the stagnation of Soviet agriculture towards the end of the Stalin era was specifically due to the defects of the collective farms.

For the yields of animal products detailed information is limited to milk and wool and figures are only available for all enterprises, State farms (excluding other State enterprises which are generally included with State farms) and collective farms. These figures are, however, completely in line with those of crop yields. The average milk yield per cow rose from 1,185 kg. in 1940 to 1,389 kg. in 1953; the increase in the State farms was from 1,803 kg. to 2,577 kg., but in the collective farms the average milk yield fell fractionally from 1,017 kg. in 1940 to 1,016 kg. thirteen years later; similarly, the wool crop per sheep rose on average from 2.2 kg. in 1940 to 2.4 kg. in 1953, with the State farms improving their return from 2.9 to 3.2 kg. and the collective farms declining from 2.5 to 2.3 kg.[27]

Only the scantiest information is available on changes in the labour resources of individual sectors. The staff of the State farms and related enterprises, excluding persons not employed

on agricultural activities, rose from 1.6m. in 1940 to 2.3m. in
1953. This increase of more than 40 per cent compares with an
expansion of 40 per cent in gross production, though labour
productivity in this sector is claimed to have increased by 6 per
cent over the same period. The average labour force of the
collective farms, again excluding non-agricultural activities, but
including the staff of the MTS which serviced the collective
farms on a contract basis, fell from 26.2m. in 1940 to 23.9m.
thirteeen years later.[28] This decline of 9 per cent is consistent
with the claimed rise of 15 per cent in the productivity of labour
on the collective farms. No official information is available on
the labour spent on the subsidiary private economy by collective
farm members (or their families), workers and employees, which
was obviously very large.[29]

In public agriculture with its heavy emphasis on large-scale
crop cultivation there is a close relationship between labour
needs and capital investment, particularly in plant and
machinery. Official index figures of the changes in the 'productive capital fund' of agriculture are as follows:

TABLE 27

The productive capital fund of agriculture 1940–53
(end of year in comparable prices, 1940=100)

	1950	1953
All agriculture	101	129
State farms, etc.	105	138
Collective farms	111	142
MTS and repair stations	129	209

Source: Selkhoz, p. 385.

These figures ignore wear and tear and may therefore be more
useful as a rough guide to the additional gross investment in
individual sectors than a comparison of the value of the productive assets employed at any moment, because the omission of
depreciation distorts the picture progressively with the effluxion
of time. They include the value of livestock, which was lower
during the post-war period than before the war, presumably
because of the sharp decline in the number of horses; otherwise
it would be difficult to account for the fact that the productive
fund of agriculture, excluding the value of livestock, rose by

147

47 per cent between 1940 and 1953 compared with only 29 per cent if livestock is included.

At the end of 1950, the gross capital employed in agriculture barely exceeded the pre-war level and the ravages of war may, on balance, just have been made good; however, those of time are ignored in this calculation and were felt in over-age and inefficient buildings and equipment, and the productivity of agricultural capital must have been appreciably less than before the war. Three years later the gross value of agricultural capital had risen by over a quarter. Direct investment in State and collective farms seems to have proceeded broadly in proportion, but a much higher share of the total was accounted for by the Machine and Tractor Stations. As the rate of capital investment in all sectors of the public agricultural system was much higher than that for agriculture as a whole, the experience of the private sector must have been much less favourable, partly probably due to the collectivization of the Baltic republics and Moldavia in the meantime. It is, however, impossible to decide whether this meant an actual drop in its productive resources or only a very much smaller rate of increase than in public agriculture.

Such abstract index figures are at best problematical; they can be supplemented by a few data on the availability of heavy equipment on State agricultural enterprises and collective farms before the war and at the end of the Stalin era (see Table 28).

Mechanization, as measured very crudely by the availability of heavy equipment for cropping, was consistently higher in the State farms (including ancillary State enterprises) than in the collective farms, but the rate of progress in the latter between 1940 and 1953 had been somewhat faster than in the former. This could be a contributory reason for the higher increase in the productivity of labour in the collective farms and does not suggest that they received a falling share of the investment funds made available to agriculture, however insufficient these may have been in total.

Market Supplies and Food Consumption

At the time of Stalin's death the war had been over for almost eight years and the economy as a whole had made considerable progress, but the growing and increasingly prosperous town

TABLE 28

Selected plant resources of public agriculture, 1940 and 1953

	1940			1953		
	State farms, etc.	Coll.* farms	Total†	State farms, etc.	Coll.* farms	Total†
1. Equipment in '000 units						
Tractors—						
physical units	87	439	531	129	615	744
15-h.p. units	118	562	684	229	1,009	1,239
Cereal combines	28	153	181	53	265	318
Lorries	29	147	228	53	241	424
2. Sown area in m. ha.						
Total	13.3	117.7		18.2	132.0	
Cereals	8.6	91.0		9.3	95.6	
3. Equipment in relation to '000 ha. sown area—units						
Tractors—						
15-h.p. to total	8.9	4.8		12.6	7.6	
Combines to cereals	3.3	1.7		5.7	2.8	

*Includes MTS.   †Includes service organizations.

Source: Section 1 Selkhoz, pp, 46, 409f, 413. Section 2 Selkhoz, pp. 46, 56. Section 3 calculated from Sections 1 and 2.

population still found it very difficult to transform its rising incomes into more and better food. Even on the basis of the official statistics, agricultural market supplies as a whole per head of the urban population in 1953 were actually lower than in the year preceding the German attack. As supplies of some important non-food items, such as raw cotton, had risen quite sharply, this must have applied with even greater force to market supplies of foodstuffs:

TABLE 29

Global supplies and per head supplies in 1953 (1940=100)

Total population 98.5

Urban population 132.8

		Per head of the population	
	Global	total	urban
Agricultural gross production	104	105.6	
crops	96	97.5	
animal products	124	125.9	
Agricultural market supplies	115		86.6
crops	110		83.0
animal products	125		93.9

Source: Narkhoz 1962, pp. 8, 227, 230 and calculations therefrom.

For a less abstract and therefore more meaningful picture it is necessary to turn to individual commodities and to their composition by sectors:

TABLE 30

Farm use and market supplies of selected products, 1940–53
(m. tons except for eggs which are '000m. units)

| | Total | | | | Public | | | | Private | | | |
| | Farm Use | | Market | | Farm Use | | Market | | Farm Use | | Market | |
	1940	1953	1940	1953	1940	1953	1940	1953	1940	1953	1940	1953
Grain	57.2	46.7	38.3	35.8	46.6	45.1	37.1	35.1	10.6	1.6	1.2	0.7
Cotton	—	—	2.2	3.8	—	—	2.2	3.8	—	—	—	—
Sugar-beet	0.6	0.3	17.4	22.9	0.6	0.2	16.3	22.9	—	0.1	1.1	
Potatoes	63.0	60.0	12.9	12.1	20.9	16.3	5.9	4.2	42.1	44.2	7.0	7.9
Vegetables	7.6	6.3	6.1	5.1	2.2	2.0	5.0	3.9	5.4	4.3	1.1	1.2
Meat and fat*	3.3	4.0	4.2	5.4	0.2	0.5	1.9	4.0	3.1	3.5	2.3	1.4
Milk	22.8	22.8	10.8	13.7	2.3	3.5	5.3	8.6	20.5	19.3	5.5	5.1
Eggs	7.5	10.3	4.7	5.8	0.4	0.7	0.3	1.8	7.1	9.6	4.4	4.0

* Live weight.
Source: Selkhoz, pp. 86f, 202f, 329 and passim.

Between 1940 and 1953 the total market supplies of such basic food crops as grains, potatoes and vegetables actually declined, primarily because of lower supplies from public agriculture, but certain technical crops such as sugar-beet and particularly cotton—both concentrated in the hands of public agriculture—rose substantially. Market supplies of livestock products, on the other hand, expanded and this applied to all three main commodity groups or commodities, meat, milk and eggs. These increases were entirely attributable to public agriculture, for the quantities coming from private agriculture fell for all three groups.

Though market supplies are in general a reasonable first approximation to the food supplies of the urban population, they were less representative during this period than in more normal times, because many town workers still kept their wartime allotments of potato and vegetable patches and also kept small livestock. To this extent it is incorrect to speak of 'farm use' as opposed to market supplies, and this factor may help to explain the apparent anomaly that the direct use of potatoes, meat and eggs in the private agricultural sector in 1953 was actually higher than in 1940 and that of milk only marginally lower, despite the completion of collectivization and the fall in the rural population.

To the extent that market supplies were thus supplemented by the self-supply of the towns, the consumption level of the urban population in 1953 may have been less unsatisfactory than the trend of market supplies might indicate; it is, however, difficult to credit as representative the official findings of workers' family budgets which claim increases of 66 per cent in the per head consumption of meat and fat, 70 per cent in that of milk products, 48 per cent in that of eggs and 96 per cent in that of sugar for 1953 over 1940.[30] Part of the fall in market supplies of some important foodstuffs, particularly from private holdings, may well be attributed to the backyard allotments of the urban population, but the dominant reason for their unsatisfactory trend was undoubtedly the low level of producer prices.

Soviet agricultural policy at the end of the Stalin era was thus faced with a baffling situation. On the production side, the greatest weakness was the stubborn failure of grain output to keep step with growing needs and a corresponding threat to the feed base of the steadily increasing livestock herds. At the same time, the continuing concentration of animal production in the subsidiary private economy of the peasants, who consumed the major part of their output themselves, kept the supply of meat, milk and eggs to the towns unduly low and increased the enforced demand of the urban population for bread and potatoes. The link between these disturbing developments was the failure of the collective farms, which remained the main force of public agriculture, to perform their basic tasks even moderately well.

3. REAL PROBLEMS AND SHAM SOLUTIONS

1928 and 1945—Parallels and Differences

On a superficial view, the German invasion and the war effort to repel it could be considered as having put the clock back to 1928, with industrialization again the overriding need of the hour, and the peasants again cast for the part of supporting this operation by supplying the necessary funds for the process of original (or repeated) socialist accumulation.

This was, indeed, the official line of the leadership at the time:

'The basic laws of increasing socialist reproduction are also essential for the post-war reconstruction and development of the national economy of the USSR. They indicate the necessity of urgent and more rapid reconstruction and development of the metallurgical, fuel and electric power industries, of the railroad transportation system . . . and also of the domestic machine-building industry. . . . The transition from a war-time to a peace-time economy is being carried out by increasing the share of accumulation in national income without which the reconstruction of the economy and its accelerated growth are impossible. This means that the share of the social product going for accumulation and reproduction is being increased, replacing military expenditures. . . .'[31]

Though, according to later claims, the author of this analysis was purged for showing undue consideration of peasant interests, his emphasis on the overriding priority for heavy industry and his quite perfunctory references to the future needs of agriculture suggest that on this occasion he successfully disguised such sentiments.

The guiding principles of Soviet policy between the end of the war and Stalin's death can be distilled in a few simple propositions, at least as far as they affect agriculture. Economic recovery required the concentration of all efforts on the reconstruction and expansion of all branches of basic industry. Consumer goods would, therefore, remain in short supply, and the village would have to provide the food and raw materials needed by the towns and the consumer goods industries, with only an insufficient supply of industrial products in exchange. Agricultural market production would thus have to proceed as far as possible not on the basis of an exchange of equivalents but on that of taxation extracted by administrative means in the form of quotas at nominal prices, obtained in the first place from public agriculture.

The reconstructed heavy industry would in due course step up the supply of agricultural machinery to public agriculture, State and collective farms, and the MTS, thus enabling it to increase output and market supplies, while releasing labour for the growing needs of industry. As soon as practicable, heavy industry would also turn out new plants for the consumer goods industries which could then expand production for the benefit of the peasants as well as of the town population. Meanwhile the procurement prices paid by the Government for agricultural supplies would have to be kept low, and the selling prices of industrial goods both for productive and for consumptive purposes, correspondingly high, in order to keep the imbalance between demand and supply within tolerable limits.

Whether the salvaging of mines and oil-wells and the reconversion of war industries to peace-time uses was, perhaps, less difficult than expected or whether as a result of concentrating all energies on this task, its progress was very fast, the pre-war level of output in most basic industries was surpassed within a few years after the end of the war.

TABLE 31
Post-war recovery in selected industries

		Gross production in			1950 as % of	Recovery
	unit	1940	1945	1950	1940	Year
Coal	m. tons	165.9	149.3	261.1	157	1947
Oil	m. tons	31.1	19.4	37.9	122	1949
Electric power	'000m. kwh	48.3	43.3	91.2	189	1946
Steel	m. tons	18.3	12.3	27.3	149	1948
Sulphuric acid	m. tons	1.6	0.8	2.1	131	1949
Metal cutting tools	'000	58.4	38.4	70.6	121	1948
Metal working tools	'000 tons	23.7	26.9	111.2	469	1945
Trucks and cars	'000	145.4	74.7	362.9	250	1948
Paper	m. tons	0.8	0.3	1.2	147	1949
Cement	m. tons	5.7	1.8	10.2	179	1948
Bricks	'000 m.	7.5	2.0	10.2	136	1949
Cotton fabrics	'000 m m^2	2.7	1.1	2.7	102	1950
Linen fabrics	m. m^2	268.3	97.9	257.4	96	1951
Stockings	m. pairs	485.4	91.0	472.7	98	1951
Leather shoes	m. pairs	211.0	63.1	203.0	96	1951

Sources: Narkhoz 1964, pp. 130ff. Jasny *Soviet Industrialization,* 1928-52 (1961), p. 368.

1940 was not a normal year, because the concentration on heavy industry, and particularly on war preparations, was already intense. Compared with this situation, the pattern of industrial output in 1950 shows a further distortion in the direction of the basic industries, while typical consumer goods industries had barely climbed back to their depressed pre-war level. The index of industrial gross production in 1945 (1940 = 100) was 92, with producer goods industries standing at 112 and consumer goods industries at 59. With the cessation or reduction of war production, output in the producer goods industries declined in 1946 and did not regain the 1940 level until 1947, but in 1950 it was over twice and in 1953 almost three times the 1940 figure; the index for consumer goods industries, on the other hand, exceeded the 1940 level only in 1949 and reached 123 in 1950 and 177 in 1953.[32]

The slow recovery of the consumer goods industries was itself connected with the backwardness of agricultural production, which in 1950 was still marginally below the 1940 figure and continued to hover around this level for the next three years. This reflected the fact that the consequences of the Second World War for the Russian village were much more severe than for industry and proved much more difficult to make good. The 1928 approach was, therefore, entirely inappropriate in 1945, however inevitable for a bureaucratic dictatorship, and however attractive for the government of a country which found itself almost by default in the position of the second super-power. In reality it was not industry which needed another 'shot in the arm' through the transfer of funds from the peasants, but agriculture in all its sectors and branches which was still reeling from the repercussions of the war on its debilitated structure. Only a massive blood transfusion in the form of lavish capital investment in its ramshackle large-scale public enterprises and higher incomes for the peasants could have enabled it to meet the needs of the progressively industrialized post-war Soviet society.

In economic terms, the most important aspect of the Second World War for the Soviet Union was not the grave setback to the material level of life and production which it entailed, nor even the ghastly loss of life amongst the most productive age groups, but the further intensification of the imbalance between heavy industry and the rest of the economy, and above all agriculture. The excesses of the collectivization crisis may be explained, though not justified, by the fact that the régime was caught up in a power struggle for survival with the peasants. By 1940 the country had barely recovered from the effects of this cataclysm on the standard of living of the population and on the productive capacity of agriculture on which half the Soviet people still depended for their livelihood.

In 1945 the power struggle between the Government and the peasants was to all intents and purposes a thing of the past, but the wilful blindness of the ruling bureaucracy to the true needs of the situation prevented real progress towards the solution of the agricultural problem during the remainder of Stalin's reign and created lasting, and still only partly resolved, difficulties for the development of the Soviet economy as a whole.

Low Farm Prices and their Consequences

The attempt to stop the clock in the relations between the régime and the peasants was nowhere more blatant than in the farm price policy pursued after the war. Information on the actual prices received by producers for agricultural products during this period is scarce but not completely lacking. As before the war, the produce of the State farms was simply transferred to the public trading system at certain accounting prices irrespective of the financial results of the individual enterprise, while by far the larger proportion of the marketable surplus of the collective farms was 'procured' by the Government and its organs; the compulsory delivery quota was acquired at almost nominal rates, while excess purchases were paid for at considerably higher prices and often in exchange for rare industrial products, and only the remainder could be sold at the high prices prevailing in the collective farm market. Output from individual plots was also subject to fairly heavy procurement quotas, but the major part of their marketable surplus was sold at the high collective farm market prices.

Typical figures for the prices paid by the Government for the obligatory quotas from a Ukranian farm in 1949 were 9.5 old* rubles per 100 kg. of wheat, 7.5 rubles for rye, 6 rubles for barley and 8 rubles for maize.[33] Compulsory deliveries from the

* The currency reform of 1961 defined the 'new' ruble as equal to 10 rubles of the earlier issues. Prices expressed in 'old' pre-1961 rubles, are therefore, ten times as high as in 'new' rubles.

peasants' private plots were paid for at the same prices as deliveries from the collective farm, but in this case cereal prices were of little importance, as the peasants produced mainly potatoes, vegetables and livestock products. For cucumbers, the Government paid 2.5 rubles per 100 kg., for cabbage 3 rubles and for tomatoes 4 rubles; potatoes also fetched 4 rubles, beef 9 rubles, pork 14 rubles and milk 3 to 5 rubles per 100 litres according to the butter-fat content, while eggs were priced at 1½ rubles per hundred.[34] As for deliveries to public trade in excess of the obligatory quotas, payment was 20 rubles per 100 litres of milk and 10 rubles per 100 eggs; such supplementary deliveries were a condition of purchase for essential but desperately short consumer goods such as kerosene, soap or window glass and others.[35]

In view of the discrepancy between the prices paid for the same commodity according to type of trade, the calculation of average producer prices paid by the Government presents obvious problems. Such price calculations for 1952 formed the basis of later publications of average producer prices in the Khrushchev era, and the absolute figures were quoted by Khrushchev himself in a speech in February 1964. In the following table they are quoted in terms of the old rubles in force in 1952 and in wheat-parity prices:

TABLE 32

Government procurement prices in 1952

(old rubles per 100 kg.; for eggs in terms of 100)

	rubles	wheat=100
Wheat	9.7	100
Maize (grain)	5.4	56
Peas	13.1	135
Green beans	14.7	152
Sugar-beet	10.5	108
Raw cotton	318.8	3,287
Sunflower seed	19.2	198
Tobacco	720.3	7,426
Potatoes	4.7	48
Vegetables	19.2	198
Beef (live weight)	20.3	209
Pigmeat (live weight)	67.2	693
Milk	25.2	260
Eggs	19.9	205

Source: Khrushchev, *op. cit.* (in III, 11), Vol. VIII, p. 467.

It is notoriously difficult to find valid standards of comparison of the relative prices of agricultural commodities in different countries or at different periods. Two alternative methods are comparisons of prices and costs of production and changes in time of prices received by farmers and those paid by them. Neither can be applied really satisfactorily, but some indications can be obtained from the existing fragmentary data.

Though comprehensive calculations of costs of production in public agriculture have been published, at least in the form of average figures, for recent years, no such figures are available for the later Stalin era, and it is doubtful whether they were available even to the authorities. Nevertheless, it is plain at first sight that most technical crops, particularly cotton and sugar-beet, were overvalued in relation to grain and that animal products were absurdly undervalued by any standard.

The relatively favourable treatment meted out to the producers of technical crops was the direct result of the dependence of much of the consumer goods industries, particularly the light and the food industry, on agricultural raw materials. The price of raw cotton was raised sharply soon after the war, and cotton-producing collective farms were assured of priority supplies of essential foodstuffs and industrial goods on favourable terms—an arrangement which Stalin regarded as a model for the relationship between the peasants and the State in general.[36]

In the case of sugar-beet, perhaps the strongest inducement was the practice of paying the suppliers partly in kind, but there was also a fairly steep price premium on above-average yields. Thus in one case, 5 rubles per hundred kg. was paid for the planned quantity which may have been 10 tons per hectare, but for the next 20 quintals per hectare the payment rose to 10 rubles and still higher yields attracted even higher prices.[37]

It is more difficult to understand the failure to pay adequate prices for animal products which were of at least equal importance for the nutrition of the urban population. It is not impossible that it was to some extent the result of ignorance of the facts on the part of the decision makers, but it is more likely that it was a reaction to the major part played by the private economy of the peasants in the production and marketing of meat, milk and eggs. Low procurement prices for animal products constituted an additional tax on the subsidiary plots, but they also made the animal husbandry of the collective farms thoroughly unremunerative and helped to depress the payment of their members for labour days. At the same time they ensured the chronic insufficiency of market supplies in face of a steadily more clamorous demand.

While producer prices of most foodstuffs paid by the Government remained at or near their pre-war levels, retail prices in public trade were sharply higher:

TABLE 33

Retail prices of selected foodstuffs in Moscow 1937–52

Item	unit	in rubles			in % of 1937	
		1937	1948	1952	1948	1952
Rye bread	kg.	0.85	3.00	1.50	353	176
Wheat bread	kg.	1.70	4.40	2.00	259	118
Potatoes	kg.	0.40	1.00	0.90	250	225
Cabbage	kg.	0.30	1.00	0.85	333	283
Sunflower oil	kg.	14.84	30.00	21.60	202	146
Beef, 1st grade	kg.	7.75	30.00	14.80	387	191
Pork	kg.	10.38	48.00	23.70	462	228
Milk	litre	1.60	4.00	2.90	250	182
Butter	kg.	17.50	67.14	30.60	384	175
Eggs	10	6.13	16.00	10.40	261	170

Source: Janet G. Chapman, *Real Wages in Russia since 1928* (1963) pp. 55ff, 190ff.

Three years after the end of the war, while Russia was in the grips of a severe food shortage, the prices of the most elementary necessities were a multiple of their pre-war level, and as far as public trade was concerned this benefited the producers much less than the Exchequer, which obtained a large proportion of the increased consumer price in the form of higher turnover taxes, though processing and distribution costs had also no doubt gone up.

The only market sector in which the changes in the balance of supply and demand affected both producer and consumer prices was the collective farm market and the authorities relied to some extent on it for mopping up some of the excessive purchasing power of the urban population. In 1948, the average price level in this market was estimated at three and a half times that of 1937, but by 1952 prices had considerably receded. As this market was, by definition, based on direct transactions between producers and consumers, it transferred excess purchasing power from the workers to the peasants which was, if anything, even less desirable. This problem was at least alleviated by the currency reform of December 1947, which sharply devalued the cash resources in the hands of the peasants and thereby cut their purchasing power for industrial goods in short supply. Furthermore, the prices in the collective farm market tended to follow at a distance those of the State shops and, therefore, gradually fell back.

This was in some ways a success of the Government *vis-à-vis* the peasants, but it was also the final link in a vicious circle. For the collective farms it reduced the extra income available for the payment of labour days and thereby weakened the already feeble material incentives for work in the collective enterprise. For the peasants in their private capacity it reduced the attraction of the only worth-while market for their produce and acted as a brake on market supplies from this sector, which was a vital source of potatoes, vegetables and all kinds of animal products for the towns.

The clash of views within the Communist leadership on the issue of agricultural producer prices broke surface for the first time in Stalin's last public pronouncement. In a scholastic discussion of the rôle of the law of value under socialism, Stalin expressly criticized a proposal by 'our business executives and planners' to fix the price of grain 'practically' at the same level as that of cotton; the actual price ratio was 1 to 33. He also accused them, less improbably, of wanting to establish a closer relationship between the price of grain and the price of bread: 'The Central Committee was, therefore, obliged to take the matter into its own hands and to lower the price of grain and raise the price of cotton.'[38]

Three or four years afterwards Khrushchev revealed an abortive attempt, shortly before Stalin's death, to improve the insufficient animal husbandry of public agriculture. A Commission was charged with drafting a resolution on the subject and proposed increased producer prices. According to Khrushchev, this proposal was vetoed by Stalin personally who suggested instead a swingeing increase in agricultural taxation.[39]

At a time when it was neither particularly dangerous nor particularly useful to do so, the official producer pricing policy of the Stalin era was also severely taken to task by S. G. Strumilin, in his day one of the most influential planners of the Soviet establishment. Six years after Stalin's death he warned that 'the practice of fixing grain delivery and procurement prices up to 1953 teaches us that serious disproportions and delays in economic development are caused by blunders in planning in this field.'[40]

The other measure of the adequacy of producer prices is their purchasing power as regards the goods that peasants wish to buy Though detailed evidence on retail prices is available only for Moscow, where they were lower than in the countryside, the evidence suggests that they had gone up, if anything, even more than the retail prices of food:

TABLE 34
Retail prices of selected articles in Moscow, 1937–52

Item	unit	rubles			in % of 1937	
		1937	1948	1952	1948	1952
Sugar, granulated	kg.	3.80	13.50	10.35	355	272
Salt, ground	kg.	0.11	1.50	0.35	1,364	318
Tea	kg.	80.00	160.00	104.00	200	130
Vodka, 40%	½ ltr.	6.55	42.75	21.60	653	330
Calico	mtr.	3.43	10.10	8.60	294	251
Cotton shirt	1	42.90	89	77.50	207	181
Leather boots	pair	102	247	210	242	206
Felt boots	pair	75	195	146	260	195
Soap, household	kg.	3.10	13.00	6.65	419	214
Kerosene	litre	0.47	2.00	1.40	426	298
Matches	box	0.024	0.20	0.12	833	500
Makhorka	packet	0.75	2.30	1.40	307	187

Source: As Table 33.

Though the close control of the Government over the main distribution channels was an invaluable adjunct of its retail price policy, the latter seems to have been no less shrewd than tough and no less flexible than determined. Nevertheless, the rise in consumer prices and the refusal of the authorities to raise official procurement prices reduced the real value of the bulk of agricultural market produce so much that it defeated the purpose of the policy as a whole.

Bureaucratic Cures for Social Ills

Towards the end of the Stalin era repeated attempts were made to deal with the symptoms of the failure of the collective farms as productive enterprises by grandiose and characteristically bureaucratic schemes.

Soon after the war, in September 1946, the Government set up a special Council for Collective Farm Affairs. Its function was the co-ordination and supervision of the work of the various departments concerned, which was complicated by the division of responsibility between the Ministry of Agriculture and the MTS and by friction between the local organs of the Ministry and those of the Communist Party. Its head was Andreyev, apparently a close rival of Malenkov, whose handling of the collective farms came under fierce criticism a few years later.

In 1948 the Communists launched the 'majestic' Stalin Plan for the Transformation of Nature in connection with the general introduction of the grassland system of management advocated by V. R. Williams. Its most spectacular elements were the plans to improve the climate of the country through the planting of tree belts on a large scale and those for a substantial expansion of the area under irrigation. The difficulties which these plans were designed to overcome were real enough. What frustrated the more ambitious—and in the long run beneficial and even necessary—of these proposals was the double failure of the régime to provide the necessary funds and to appreciate that such plans could only be carried out after the gross exploitation of the peasants by the State had come to an end. Otherwise they were bound to remain on paper, and their intrinsic merits were compromised by the bureaucractic half-measures adopted for carrying them out. Thus it is not surprising that some elements of this policy were dropped quickly after Stalin's death—to be replaced by other but not necessarily more successful panaceas —while the 'struggle' against the grassland system of management lasted until almost the end of the Khrushchev era.

Of more permanent importance was the policy of amalgamating small collective farms into larger units. This was officially put into operation in 1949-50, soon after the elimination of the last remaining strongholds of individual peasant farming in the territories acquired during the Second World War.

One of the strongest political incentives for this policy was the weakness of the Party units in the countryside and their complete absence in many collective farms. The most important argument of a strictly economic character was the fact that many of the then existing collective farms were too small to make use of the economies of scale obtainable from the desperately scarce resources of skilled management and mechanical traction.

Collectivization was in the beginning generally based on the single village: a number of mixed dwarf holdings were merged into a single unit which often continued to be more of an agglomeration of its original elements than an organized enterprise. Economies of scale in the use of mechanized equipment were obtained by concentrating most farm machinery in the MTS, but the collectivized village remained a poor basis for large-scale production and an even worse one for creating the badly needed social and economic infrastructure. The principles underlying the amalgamation policy were, therefore, fundamentally sound, but its practice showed the customary defects of a blunt and insensitive bureaucracy at work.

The Party and Government machine went resolutely into action. In 1940, the number of collective farms amounted to 235,000; in 1950, after the new policy had been in operation for a single year, and despite the collectivization of important new territories, only 121,400 collective farms were left, and three years later these had been further reduced to only 91,200.[41] The practical problems caused by such sudden expansion of the area of activities of individual collective farm units must have been very great. Before the war, the average farm contained 81 households with a sown area of 492 hectares; this may well have remained a reasonable measure of the dimension of collective farms until 1949, when it amounted to 557 hectares.[42] By 1953, the average number of households had increased to 220 with a sown area of 1,407 hectares. Common livestock holdings had gone up more than proportionately, with almost 300 head of cattle (including 93 cows), 146 pigs and over 800 sheep and goats per farm. To carry out a reorganization of hundreds of thousands of enterprises by bureaucratic *fiat* within a few years must have disturbed rather than improved farm operations and reduced rather than increased the efficiency with which they were carried out.

Amalgamation inevitably increased the administrative disadvantages flowing from the dispersion of the rural population into very small villages or settlements. Ten years later, in 1961, there were over 700,000 of them; half contained fewer than 25 persons with only 2.2 per cent of the population and only 7.4 per cent consisted of more than 500 people, though more than half the rural population lived in them.[43]

The proposal, first made by Khrushchev in 1950, to regroup the collective farm peasants into 'agro-towns' consisting of newly-built houses, or even flats most favourably situated in relation to the amalgamated collective farms, must be seen against this background. One of the obstacles in the way of the technical reconstruction of the collective farm economy through amalgamation was the location of the subsidiary private plots. As a rule they were found in the immediate vicinity of the peasant dwelling and the peasants spent as much time as possible on them, and particularly on their livestock which needed daily attention. The new crop rotation system made the fodder supply for private livestock even more precarious and amalgamation increased the distance between home and work on the reorganized fields. This made it even more difficult for them to combine collective and private work, and as their income from the produce of their private plots was generally more important to them than that from their 'labour days' in the collective farm, it was the latter that tended to suffer most.

One of the consequences, if not the purpose, of the planned 'agro-towns' would have been to separate the peasants from their cherished private plots: 'In forming new settlements and also in the reorganization of the old villages a large personal allotment should not be laid out near the house, since in this case the village would occupy a very large area. . . . Proposals have been made to restrict the personal allotments in the settlement to small proportions of 0.1 to 0.15 hectares. . . . The remainder of the area of personal allotments stipulated by the Collective

Farm Statutes would be removed beyond the settlement limits. . . .'[44]

The moves which preceded the rejection of this bold proposal are a matter of conjecture, though Stalin and Malenkov are known to have played a decisive part in them;[45] whether it was the huge cost of the scheme and the anticipated resistance of the peasants or the tug-of-war between factions in the Kremlin, the plan was not seriously pursued, though its rejection did not cause any lasting damage to its versatile author.

The lively discussion on the future of the collective farms continued within the Communist Party until Stalin's death. The last article in the collection of his pronouncements *ex cathedra*, published late in 1952, indicated the strength of the two opposing tendencies within the Soviet High Command on this subject. On one side were the advocates of transforming the collective farms into State farms; on the other, some economists proposed that, on the contrary, the State should wash its hands of the heavy responsibility for the capital outlay needed for the modernization of the collective farms and should transfer the MTS to them. This conflict was played out, with paradoxical results, in the Khrushchev era.

At the moment, however, it was Stalin's view that was decisive and he was opposed to both extremes. His objection to the nationalization of the collective farms was, of course, not on grounds of principle but for reasons of timing and methods; he admitted that 'group or collective farm property and commodity circulation' were for the time being appropriate and beneficial but criticized the 'unpardonable blindness' of ignoring that they were already beginning to hamper the development of the productive resources of the country, 'since they create obstacles to the full extension of Government planning to the whole of the national economy, especially agriculture'. He concluded that the main task for the future was that of 'gradually converting collective farm property into public property' and of including surplus collective farm output 'in the system of products exchange between State industry and the collective farms',[46] in other words abolition of the collective farm market and its replacement by a system of contracts for the whole marketable produce of the collective farms to public trade.

164

Needless to say, Stalin disagreed with the proposal of Sanina and Venzher to transfer the MTS to the collective farms. In his view, the MTS were decisive for 'the fate of agriculture in present-day conditions'. His main objection was, therefore, political because it was based on considerations of power. Nevertheless, he was also economically on strong ground when he doubted whether the collective farms could invest the huge sums needed to expand and modernize the agricultural machinery of the country and concluded that 'the effect of selling the MTS to the collective farms as their property would be to involve them in heavy loss and to ruin them, to undermine the mechanization of agriculture'.[47] Though over-drawn and high-coloured, this picture was not quite imaginary, and the experience of the collective farms in the years immediately following their compulsory acquisition of the MTS in 1958 was certainly far from happy.

Despite their dogmatic form, Stalin's last thoughts reflected faithfully enough the unsolved problems in the relations between the Soviet power and the peasants at the time—and during the years ahead.

VIII

The Khrushchev Era
(1953-1964)

1. THE TASK BEFORE THE REFORMERS

The ossified rule of an absolute dictatorship had led to a number of grave difficulties for the Soviet régime in various fields. The Cold War with the United States and its military, diplomatic and economic consequences, the acute conflict with Yugoslavia and the growing tension between the Soviet Union and its Eastern European satellites all had to be tackled sooner rather than later. The most serious domestic problem confronting Stalin's successors was, however, the need to forge a satisfactory relationship with the peasants, and the agricultural dilemma became one of the most potent sources of friction between the new rulers.

The agricultural system was patently incapable of meeting the growing claims of Soviet society and in need of radical reform. The open power struggle between the peasants and the régime had, of course, been decided once and for all, and the Collective Farm Statute might have become the basis of a tolerable compromise, if it had been allowed to operate in practice: it would have freed the towns from the threat of being held to ransom under the threat of starvation and permitted the peasants to keep an acceptable proportion of the benefits of rising output for themselves.

This proportion was undoubtedly much lower than would have satisfied the farming community of any advanced industrial Western country, where the economic protection of agriculture has become the rule rather than the exception. The actual economic position of the farmers in these countries, however unsatisfactory it may appear to them, is thus better than it would be in 'free market conditions'. In Soviet Russia, on the

166

other hand, this had not been the case even in the years of the New Economic Policy, to which the peasants of the next generation might have looked back as to a Golden Age. Even then, the keenest mind amongst the Marxist economists of the day, Y. Preobrazhenski, was gravely concerned about the crying inequality between the value of farm produce marketed by the peasants and that of the industrial products available to the village in exchange, when he warned that 'the law of value' might reassert itself with destructive force in the relations between the régime and the peasants.

After the Second World War the same question occupied the authorities both in theory and in practice. Its ghost haunted Stalin's last theoretical utterance shortly before his death. He blandly repeated Lenin's statement[1] that the 'antithesis between town and country' was rooted in the exploitation of the country by the town and therefore fated to disappear with the abolition of capitalism, and added flatly, with a grim humour all his own, 'and that is what happened'.[2] (A year earlier, his then henchman Khrushchev had only gone so far as to claim that the process of eliminating this antithesis was making gradual progress.[3])

At the same time Stalin accepted the fact that, even after collectivization, the peasants were 'unwilling to alienate their products except in the form of commodity exchange for which they desire to receive the commodities they need. At present collective farms will not recognize any other economic relation with the town except the commodity relation . . . and because of this, commodity production and trade are as much a necessity with us as they were thirty years ago. . . .'[4] Unfortunately for the peasants—and for the peace of mind of Stalin's successors —the reality of the 'commodity exchange between town and country' was at the time of his death, as throughout Russian history, a caricature of the Marxist definition of this category, which was supposed to be based on the equivalence of socially necessary labour embodied in the products exchanged. This was the root cause of the failure of market supplies to expand in line with the voracious needs of the towns; it compelled the new leadership, for the first time since the 1920s, to rethink their agricultural policies.

It is doubtful whether Khrushchev's rise to the position of first secretary of the Communist Party, as distinct from Stalin's place as general secretary, owed anything to his prominent but not invariably successful part in the direction of agricultural policy during the preceding years. However, once arrived at the most senior place in the party hierarchy, he identified himself with all his exuberant energy from the beginning of the new era with distinctive policies on all the main agricultural issues and used his command of the Party machine with great determination in order to put these policies into practice.

The problem was being tackled on a very broad front. A semi-official statement lists the main measures as the re-establishment of the principle of material incentives for the agricultural population in the expansion of output, the strengthening of the administrative and technical staffs of collective farms and State farms, the removal of bureaucratic distortions in agricultural planning, increased capital investment, the expansion of grain cultivation in the 'new lands' and, as its direct result, the specialization of collective farm operations, the absorption of the MTS by the collective farms and the unification of producer price-levels for deliveries to the Government.[5]

This is little more than a recital of the measures actually undertaken by the régime between 1953 and 1958, and the hopeful claim of the learned authors that the application of such policies had accomplished the 'liquidation of the backwardness of agriculture' did more credit to their orthodoxy at the time of writing than to their understanding. Nevertheless, the changes made during the period 1953 to 1958 were so far-reaching that these years rank with the great collectivization crisis as the most dramatic era of Soviet agricultural policy. However, the early Khrushchev period differs from the early 1930s not only in its freedom from brutality and terror and the infinitely greater understanding of the economic relations, as distinct from the power questions, involved, but also in the quite substantial successes achieved by the policy makers in raising the volume and the efficiency of agricultural production.

So much progress was made during these years that the Soviet leaders were tempted to proclaim as a short-time objective the aim of overtaking the United States as producers of butter and meat, the never-never land of Soviet economic mythology. As it happens, the decision to base future policies on the extrapolation of past trends was made public at the very moment when the impetus had been exhausted which had given rise to the earlier successes: after 1958 agricultural progress did not entirely stop, but the stubborn failure of crop production, and particularly of the still decisive cereal economy, to develop in line with the growing needs of the population undermined its essential basis, even before the severe harvest failure of 1963 engendered a state of manifest crisis which ultimately engulfed Khrushchev himself.

2. THE STRATEGY OF AGRICULTURAL GROWTH

Intensification and Extension

The upshot of the violent theoretical discussions at the end of the 1920s had been a two-pronged attack on the backwardness of the dominant peasant agriculture: collectivization of the small peasant holdings with the support of mechanization was regarded as a decisive measure of intensification of agricultural production, while simultaneously the arable land area was to be expanded through the creation of mechanized 'grain factories', owned and managed directly by the State, in the outlying provinces, particularly in western Siberia and Kazakhstan.[6]

The hopes pinned on these developments were disappointed in the event. Mechanization of the basic processes of grain cultivation certainly permitted a reduction in agricultural employment and freed the land previously used for growing the fodder of millions of horses, but in the disastrous climate of forced mass collectivization the productivity of the collective farms was deplorably low. The giant State farms, on the other hand, were mostly costly failures. The combination of intensification and extension had completely eluded the régime.

After the recovery of agricultural production from the ravages of war within the framework set by the policy of the 1930s, when grain production and particularly feed grains stagnated around the level reached in 1950, the search for a new strategy of agricultural growth became again an urgent issue. It was common ground amongst the Communist leadership that agriculture had been starved of resources since the war and that more funds had to be pumped into it, partly to improve the productive system and partly to give the peasants adequate incentives for harder and more purposeful work. Disagreement arose, however, on the question whether immediate emphasis should be placed on raising the abysmally low crop yield of the existing agricultural land or whether extra output should be obtained in the first place through an expansion in the sown area.

Higher yields meant more intensive cultivation requiring more productive inputs such as larger supplies of mineral fertilizers. For an area of about 200m. hectares an annual application of, perhaps, 70m. tons of fertilizers was needed,[7] but the supplies to agriculture in 1953 were less than 10 per cent of this quantity or some 6.6m. tons. Although this figure was stepped up to 10.6m. tons in 1958[8] it remained pathetically low in relation to total needs and mineral fertilizers were, in fact, used mainly for selected industrial crops, such as cotton and sugar-beet, where they promised the highest returns. A policy of massive fertilizer application to grain production involved the building of new plants on a huge scale which would absorb not only large capital resources but also a lot of time—and while capital was short, time was of the essence of the matter.

The 'New Lands' Campaign
The most powerful advocate of tackling the problem of insufficient grain production in the first place through an expansion of the cultivated area in the semi-arid 'virgin' lands of Siberia and Kazakhstan was Khrushchev. He proposed to exploit the accumulated soil fertility of lands which had either not been utilized at all in the past because of the great climatic hazards involved, or left untilled since the ill-starred giant grain factories of the early 1930s—'left to the rabbits and wild goats'.[9] He was in favour of taking a very great but calculated risk for the sake of a considerable possible short-term gain, the lifting of agricultural production out of the rut in which it might otherwise have remained for years.

In March 1954 Khrushchev obtained the agreement of the top leadership to a programme which involved the planting of

13m. hectares; this was doubled in August, when it was decided to sow 20-30m. ha. in two years[10]—'the same amount of land that was to have been reclaimed only after many years and at large cost under the Great Stalin Plan for Remaking Nature'.[11] According to the official statistics, but probably also in fact, this huge programme was more than fulfilled: between 1954 and 1956, 35.9m. ha. were put under the plough, including 19.9m. in Kazakhstan, 5.6m. in western Siberia, 2.1m. in eastern Siberia and the Far East, 3.3m. in the Urals and 1.4m. in the Saratov and Stalingrad districts along the Volga.[12]

The extent of the geographical and economic reorientation of Soviet grain production during the five years after Stalin's death is shown in the following table:

TABLE 35

The 'new lands' campaign 1954–58

	Total USSR		'New lands'		Rest of country	
	1953	1958	1953	1958	1953	1958
1. Sown area (m. ha.)						
wheat	48.3	66.6	20.8	43.5	27.5	23.1
all cereals	106.7	125.1	36.7	60.1	70.0	65.0
other	50.5	70.5	11.3	20.6	39.2	49.9
total	157.2	195.6	48.0	80.7	109.2	114.9
2. Gross production (m. tons)						
wheat	41.3	76.6	15.9	43.3	25.4	33.3
all cereals	82.5	141.2	26.9	58.4	55.6	82.8
3. Yield (quintals per acre)						
wheat	8.6	11.5	7.7	10.0	9.2	14.4
all cereals	7.8	11.3	7.3	9.7	7.9	12.6

Source: Selkhoz, pp. 224ff and calculations from district data.

1958 was the high-water mark of success for Khrushchev's agricultural policy, largely because it was climatically an exceptionally favourable year, though in the 'new lands' weather conditions were, in fact, not so good as they had been in 1956. The total sown area in 1958 was some 38m. ha. larger than in 1953 and this was almost entirely due to increased cultivation in the new lands. The expansion in the area under grain was, indeed, entirely concentrated on the new lands; the grain acreage in the traditional producing areas was 5m. ha. or 7 per cent lower in 1958 than five years earlier, though the bountiful harvest of

171

1958 made it possible to obtain a substantially higher crop from fewer acres.

Even more spectacular than the shift in grain production was the change in the sources of Government procurements. In 1953 the regions containing the new lands furnished just over one-third of the marketable surplus of grain; five years later they accounted for well over half of a larger total.

In 1958, the *annus mirabilis* of the Khrushchev era, not even the relatively low yields of the new lands need have held any terrors for the authorities. Grain yields in general and wheat yields in particular were, indeed, considerably lower in the new lands than in the settled areas as would be expected, if for no other reason, from the fact that they produced only spring wheat. Nevertheless, they compared well with the national averages five short years earlier; it would have been asking too much of human nature to expect that Khrushchev should have attributed these highly satisfactory results to the weather rather than to his own policy.

In the short run, and within certain limits, this policy had undoubtedly paid off, though there were bad years such as 1955 and 1957 as well as good ones such as 1956 and 1958. According to Khrushchev's calculations, which may have to be taken with a grain of salt, the Soviet exchequer showed a net profit of 3,000m. rubles on the new lands campaign between 1954 and 1962, apart from an increase of 4,700m. rubles in the productive funds of State farms and other public bodies.[13]

However, the chief benefit obtained by the development of the new lands was that it had bought time—the time needed to modernize and intensify agriculture in its traditional areas which did not suffer from such frightening climatic hazards. Properly understood, Khrushchev's dash for expansion in the virgin lands was not an alternative to the policy of intensifying agriculture in the black earth belt, the Ukraine and the northern Caucasus, but a stop-gap operation which might have made it possible to pursue this long-term policy in relative comfort rather than under the pressure of a permanent grain deficit.

In practice, the resources available were probably insufficient to tackle both tasks at the same time. Although Khrushchev claimed that the public investment in the development of the new lands from 1954 to 1959 amounted only to 37,400m. (old) rubles or not more than about one year's gross investment in agriculture, the total tractor horse-power in the rest of the Soviet Union at the end of 1958 was exactly the same as it had been at the end of 1953—852,400 tractors in 15-h.p. units; the whole net increase was concentrated in the new lands, where it rose from 386,100 to 697,800 units of 15-h.p. It is, however, true that the number of grain combine harvesters outside the new lands rose from 198,900 to 254,700 which compared with an increase in the new lands from 118,800 in 1953 to 247,000 in 1958. At the end of 1958, about 40 per cent of the tractor horse-power and almost half the grain harvesters were stationed in the new lands areas.[14] Such figures explain why it could be claimed after Khrushchev's fall, when it was politically safe to do so, that in some years the new lands obtained an excessive share of capital investment at the expense of other regions, and particularly of the central non-black-earth districts.[15]

It was psychologically and financially impossible to concentrate to the required extent on the development of the new lands, while at the same time labelling this policy as a temporary stopgap until the time when agricultural production in the traditional areas could be raised to higher levels of output through intensification. But the inevitable result of designating the virgin lands as the normal and permanent source of a high proportion of the grain resources needed by the towns was the premature exhaustion of their slender reserves of fertility, accompanied by progressive weed infestation, which led to another serious grain crisis within a few years.

These dangers had been evident from the start. They were stressed by Khrushchev's domestic opponents and by foreign experts who warned of the consequences of continuous cropping, insufficient fallow and the risk of creating a gigantic dust-bowl.[16] The official leadership ignored these threats. As late as the autumn of 1961, Khrushchev demanded the fullest possible use of the virgin lands, 'the creation of our Party, the pride of our country'. He castigated the Party workers of Kazakhstan for resorting to the monoculture of spring wheat, but advised rotations including maize, peas and other leguminous crops, sugar-beet and only 'when it was essential' clean fallow. The

173

new aim was to use the new lands as a livestock base, i.e, to treat them in every respect like the settled agricultural areas: 'The virgin lands have given our people thousands of millions of poods of grain, now they must give, in line with grain, millions of tons of meat, milk and other livestock products.'[17]

The changes in the cropping system of the older agricultural districts during this period indicate that Khrushchev and his advisers believed that the development of the new lands had created the basis of a new geographical specialization, a permanent rearrangement of output patterns, with the traditional grain-producing areas switching over more and more from bread grains to fodder crops. The major change was completed by 1958 and the first indication of a shift away from the extreme pattern established in the heyday of the new lands campaign appeared only in 1964, after the crop failure of 1963.

In the following table, changes in grain and wheat acreages are shown for broad geographical areas. The division between European Russia and Siberia does not adequately reflect the effects of the virgin lands policy, because some parts of European Russia—the Lower Volga and Urals regions—contained substantial 'new lands' districts, but the comparison is, nevertheless, instructive.

TABLE 36

Acreage under wheat and all cereals by regions, 1953–64
(m. hectares)

	Wheat				All cereals			
	1953	1958	1963	1964	1953	1958	1963	1964
European Russia	22.0	23.3	24.8	26.9	54.5	54.4	60.4	62.5
Siberia*	7.8	12.8	13.6	13.5	13.7	18.1	19.0	19.1
Kazakhstan*	4.6	19.3	18.2	18.2	7.0	23.2	24.1	24.4
Ukraine and Moldavia	10.2	8.3	5.5	6.5	21.2	17.4	17.9	18.1
North-west†	1.1	0.5	0.5	0.5	6.1	4.5	4.9	5.2
Transcaucasia	1.2	0.9	0.8	0.8	2.0	1.6	1.4	1.4
Central Asia	1.5	1.5	1.2	1.5	2.2	2.2	2.3	2.6
Soviet Union	48.4	66.6	64.6	67.9	106.7	121.4	130.0	133.3

* Main 'new lands' areas.   † White Russia and Baltic Republics.
Sources: Selkhoz, pp. 147, 150. Narkhoz 1964, pp. 279ff.

The new lands campaign increased the total wheat acreage by 18m. hectares in three or four years and concentrated the whole of the increase into Kazakhstan and the development areas of Russia, while actually reducing it in the Ukraine, the traditional main granary of the country, as well as in the other regions where it always had been of subordinate importance. One of the consequences of this change was increasing reliance on spring wheat with its lower yields and greater seasonal fluctuations; virtually the whole of the increase was, in fact, in this variety and the winter wheat acreage dropped from 37 per cent of the total in 1953 to only 27 per cent five years later.

Not only were the harvests in the new lands much less stable than in the rest of the country, the same was true of their population. Many of the recruits who went east in the course of the heavily publicized virgin lands campaign were repelled, or worn down, by the primitive living conditions and the tenuous links with civilization and drifted back to a less forbidding environment.

The Pro-maize and Anti-grassland Crusades

The foundation of the official cropping policy during the Khrushchev era was the belief that the rising demand for bread grains could be permanently met from the new lands and that wheat production elsewhere could, therefore, be safely reduced and the land released from wheat growing could be used for the creation of an adequate fodder base for the rising livestock population needed to satisfy the voracious appetite of the towns for livestock products.

'What matters in animal husbandry?' Khrushchev asked. 'Fodder. With fodder there is livestock, meat and milk. No fodder—the livestock perishes, there will be no meat nor milk. Where shall we get the fodder, what are the sources for producing it? We must look the facts in the face: if we remain with our present structure of sown areas, with our present pattern of fodder crops and our current yield, there will be no fodder, no meat nor milk either today or tomorrow.'[18] This passage reflects the essentially simple strategy in favour of maize cultivation. The ample use made of maize in the United States, for the Soviet leaders the exemplar of a successful advanced agriculture, its great advantages as a source of calories for livestock feeding and, perhaps, the fact that Khrushchev's long connection with the Ukraine must have familiarized him with maize as a crop in Soviet conditions, all contributed to the decision to expand maize cultivation wherever possible.

The length to which this campaign was pushed and the excesses which it produced, are apt illustrations of the pathology of a ruling bureaucracy, for the climatic limitations of maize cultivation in the Soviet Union are especially severe. The growing of maize as a cereal was necessarily concentrated on the Ukraine, the north Caucasus and parts of the central black earth belt, i.e, on some of the best winter wheat districts of the Soviet Union, with the result that by 1963 the wheat acreage in the Ukraine was not much more than half of what it had been in 1953. However, maize was planted for silage and green fodder in practically every part of the country, from north-western Russia and the Baltic republics to central Asia. At the height of the maize campaign, in 1962, 37m. hectares were under maize; only 7m. of this enormous acreage were harvested as ripe maize grain, another 7m. as unripe grain (milky-waxy stage) and 23m. were cut for silage and green fodder.[19] Khrushchev's insistence[20] that the planting of maize was justified in all regions of the country, even though climate prevented the harvesting of ripe maize, appears even more perverse in view of his repeated and powerful attacks on the universal claims made for the grassland system of management; in both cases the only satisfactory explanation of an apparently wilful disregard of elementary facts is the difficulty, or impossibility, of selective and discriminating action by a thoroughly bureaucratic administration.

The self-defeating search for a quick and simple remedy for the fodder shortage was not exhausted by the maize campaign; it included the official condemnation and practical discouragement of certain crops (particularly rye and oats) as low-yielding even in areas unsuitable for higher-yielding alternative crops and the expansion of leguminous crops and the growing of sugar-beet for fodder on a substantial scale but with problematical effects.[21] Neither the cultivation of maize nor that of leguminous plants was intrinsically unreasonable, though Khrushchev and his advisers almost certainly did not fully realize how great their claim on the limited resources of agriculture in energy, materials and, above all, in labour was;[22] however, official insistence on quick results coupled with inadequate resources and administrative compulsion in the execution of policy discredited these policies which in their proper place were well worth a trial.

The practical link between the maize campaign and the barrage of official criticism of the grassland system of management and the place of 'low-yielding crops' in crop rotation in general was the growing scarcity of readily exploitable land. Thus it is only natural that the climax of this policy of agricultural mass campaigns was reached during the years when the 'new lands' development programme ran into a phase of falling yields due to the exhaustion of the slender soil fertility of the land. The only other land reserve which could be utilized for growing more fodder crops was part of the area lying fallow or under grass. On balance, between 1958 and 1962 the arable area under clean fallow was reduced from 24.0m. ha. to only 7.4m. and the area under grass from 31.1m. to 27.3 m. ha., a total drop of over 20m.; this compares with a rise of 17.4m. ha. in the area under maize and of almost 3m. devoted to sugar-beet for fodder.[23]

The so-called 'administrative methods' employed for carrying out this reorientation of cropping patterns showed the bureaucratic character of the régime at its most futile: it was not simply a question of substituting a worse system for a better one by unreasonable orders from above, for in many cases the authorities actually scrapped existing rotation plans without putting any considered alternative system in their place,[24] unless the growing of maize in climatically unsuitable conditions can be dignified by that name.

Khrushchev was probably right when he described (in March 1962) the grassland rotation system as a sophisticated form of extensive cropping;[25] he was, of course, careful not to include in this condemnation the virgin lands campaign to which he owed a good deal of his own prestige. Nevertheless, his new-found enthusiasm for the intensification of agriculture, with the help of steadily more astronomical claims on the future output of the fertilizer industry, indicated that the potentialities of the new lands development policy had been exhausted and signalled a major reversion of policy. This came too late to prevent the debacle of 1963 but it was, nevertheless, an important turning point in agricultural policy.

State Farms and Collective Farms

Khrushchev's strategy of agricultural expansion involved not only changes in the geographical balance of production and the use of land, but also a major rearrangement of the main forms of public agriculture, the State farms and the collective farms.

The expansion of the State farms in the early post-war period had been solid rather than spectacular. After Stalin's death their economic importance increased dramatically because they were the inevitable instrument for the development of the new lands in Siberia and Kazakhstan. This did not at first involve a net rise in their numbers, for numerous foundations in the new territories were almost offset by reductions elsewhere, particularly in the Ukraine and the black earth belt of Russia. After 1956, however, their numbers rose rapidly as a result of the policy of converting some collective farms into State farms.

Though this policy was pursued only for a limited period, mainly from 1957 to 1962, its actual effects and potential importance warrant a closer examination of the available statistical material.

TABLE 37
Main State farm indicators, 1953–64

		1953	1956	1958	1962	1964
Number		4,857	5,098	6,002	8,570	10,078
Total staff	'000	1,844	2,168	3,835	6,893	7,268
Tractors (in 15-h.p.)	'000	165	311	536	1,049	1,209
Grain combines	'000	42	86	168	256	254
Lorries	'000	40	69	140	299	324
Sown area	m. ha.	15.2	31.5	52.5	86.7	87.3
incl. grain	m. ha.	7.8	22.4	37.1	59.6	59.7
techn. crops	m. ha.	0.4	0.6	1.4	3.0	3.2
potatoes and vegetables	m. ha.	0.3	0.5	0.9	1.7	1.9
fodder crops	m. ha.	6.6	8.0	13.1	22.4	22.5
Cattle	m.	3.4	3.8	8.2	21.0	22.2
incl. cows	m.	1.1	1.5	2.8	7.4	8.3
Pigs	m.	3.5	5.3	8.1	16.8	11.5
Sheep and goats	m.	10.1	10.8	26.4	41.2	44.2

Sources: Selkhoz, p. 46. Narkhoz 1962, p. 357. Narkhoz 1964, p. 410.

Between 1953 and 1956, with barely changed numbers and only a moderate increase in manpower, the sown area of the State farms almost doubled and the grain acreage almost trebled, while livestock herds rose for cows, but not for other cattle, and for pigs, but not for sheep and goats. This was the heroic period of new lands development which was almost entirely a matter of employing the State farms as the chosen instrument for expanding the spring wheat acreage through a crash programme of enormous dimensions.

From 1957 until 1962 the number of State farms rose rapidly from year to year but not nearly so fast as their staff, which almost doubled in the course of two years and more than trebled over the period as a whole. Mechanical equipment—tractors, combine harvesters and lorries—also trebled or more than trebled between 1956 and 1962 in line with the further rise in the sown area; however, although the grain acreage went up fast, even stronger emphasis was put in terms of proportionate expansion on technical crops, potatoes and vegetables as well as fodder crops.

The expansion of livestock during these years was even more spectacular: in 1962 the State farms contained almost six times as much cattle, almost five times as many cows, four times as many sheep and goats and three times as many pigs as in 1956. As far as the increase in the land and labour resources of the State farms during this period is concerned, it was not a net addition to the total resources employed in agriculture but the result of a reorganization involving the conversion of collective farms into State farms. This process extended, of course, also to livestock and equipment but it will be seen that the net increase in these resources was so large as to permit the collective farms to maintain or even increase their stocks in these fields, despite the conversion programme.

In many other respects, the experience of the collective farms during these years provides the natural foil to that of the State farms.

Between 1953 and 1956 the number of collective farms fell by a steady 3,000 a year; the number of peasant households rose marginally and the sown area quite substantially, while livestock herds tended to mark time. These figures reflect the con-

TABLE 38

Main collective farm indicators, 1953–64

		1953	1956	1958	1962	1964
Number	'000	91.2	83.0	67.7	39.7	37.6
Households	m.	19.7	19.9	18.8	16.3	15.9
Tractors (in 15-h.p.)*	'000	1,009	1,153	943	1,128	1,343
Grain combines*	'000	265	283	258	236	228
Lorries*	'000	165	323	382	416	459
Sown area	m. ha.	132.0	152.5	131.4	114.4	110.8
incl. grain	m. ha.	95.6	102.5	84.3	70.3	67.7
techn. crops	m. ha.	10.9	12.3	10.8	11.0	11.9
potatoes and vegetables	m. ha.	5.0	5.6	5.2	3.5	3.5
fodder crops	m. ha.	20.5	31.8	31.1	29.6	27.7
Cattle	m.	27.8	27.9	32.1	38.6	37.1
incl. cows	m.	8.7	10.8	11.5	13.6	13.7
Pigs	m.	13.6	16.2	23.2	31.9	22.3
Sheep and goats	m.	77.9	74.8	75.1	67.5	54.0

* 1953 and 1956 figures include Machine and Tractor Stations.
Sources: Selkhoz, p. 57. Narkhoz 1962, p. 330. Narkhoz 1964, p. 390.

tinuation of the policy of amalgamating collective farms into larger units, with some bias towards increased grain production and a very substantial expansion of the area devoted to fodder crops, with half the increase coming from maize for green fodder and the other half from increased leys.

During the following six years the conversion of collective farms into State farms was superimposed on the amalgamation process. The fall in numbers went on at twice the earlier rate and over three and a half million households disappeared in order to provide the man-power needed for the expansion of the staff of the State farms, which rose by almost three and three-quarter million persons. The sown area on collective farms declined by almost 40m. hectares and this huge acreage supplied the major part of the corresponding expansion by 55m. on the State farms. The bulk of the decline took place in grain cultivation, but the fall in the potato and vegetable acreage was proportionately even heavier; technical crops and fodder crops were much better maintained. On the other hand, there was an impressive rise in the number of cattle, particularly cows, and pigs, but the sheep and goat flock declined moderately.

After 1962 the decline in the numbers and activities of the collective farms was not completely halted, but its pace was much slower. The mechanical resources of the collective farms in tractors and lorries were fully maintained over the whole period and the number of cereal combine harvesters fell much less than the grain acreage; thus there was a considerable increase in mechanization per man and per hectare.

TABLE 39

The number of collective farm households, 1953–64 (in m.)

	(i) Total numbers					(ii) Changes			
	1953	1956	1958	1962	1964	1956/ 1953	1958/ 1956	1962/ 1958	1964/ 1962
Russia	9.20	9.16	8.41	6.42	6.26	−0.04	−0.75	−1.99	−0.16
Ukraine	5.48	5.58	5.57	5.33	5.27	+0.10	−0.01	−0.24	−0.06
Moldavia	0.48	0.51	0.54	0.56	0.56	+0.03	+0.03	+0.02	—
South	5.96	6.09	6.11	5.89	5.83	+0.13	+0.02	−0.22	−0.06
White Russia	1.28	1.28	1.18	0.92	0.91	—	−0.10	−0.26	−0.01
Lithuania	0.36	0.35	0.34	0.30	0.28	−0.01	−0.01	−0.04	−0.02
Latvia	0.22	0.21	0.19	0.17	0.15	−0.01	−0.02	−0.02	−0.02
Estonia	0.12	0.11	0.10	0.08	0.08	−0.01	−0.01	−0.02	—
North-west*	1.98	1.95	1.81	1.47	1.44	−0.03	−0.14	−0.34	−0.05
Georgia	0.47	0.49	0.49	0.46	0.46	+0.02	—	−0.03	—
Azerbaidjan	0.29	0.30	0.32	0.36	0.29	+0.01	+0.02	+0.04	−0.07
Armenia	0.15	0.15	0.15	0.14	0.12	—	—	−0.01	−0.02
Transcaucasia	0.91	0.94	0.96	0.96	0.87	+0.03	+0.02	—	−0.09
Uzbekistan	0.68	0.72	0.64	0.73	0.75	+0.04	−0.08	+0.09	+0.02
Kirgizistan	0.18	0.19	0.18	0.17	0.17	+0.01	−0.01	−0.01	—
Tajikistan	0.17	0.18	0.19	0.22	0.23	+0.01	+0.01	+0.03	+0.01
Turkmenistan	0.10	0.11	0.11	0.13	0.14	+0.01	—	+0.02	+0.01
Central Asia	1.13	1.20	1.12	1.25	1.29	+0.07	−0.08	+0.13	+0.04
Kazakhstan	0.54	0.55	0.42	0.27	0.22	+0.01	−0.13	−0.15	−0.05
Soviet Union†	19.74	19.89	18.83	16.25	15.89	+0.15	−1.06	−2.58	−0.36

* Including Kaliningrad *oblast*.　　† Differences due to rounding.
Sources: 1953–58: Selkhoz, p. 52. 1962: Narkhoz 1962, p. 345. 1964: Narkhoz 1964, p. 402.

One of the most important features of the conversion policy was its unequal *geographical* spread, which can be measured by the change in the number of collective farm households in different parts of the country. In the preceding table this is limited to Union Republics, though at least for the Russian Republic more detailed figures are needed for a reasoned assessment.

Within the huge Russian Federated Republic there were striking regional differences: in the lower Volga region and the north Caucasus the number of collective farm households changed on balance hardly at all between 1953 and 1964, while in the Urals and Siberia the fall exceeded 50 per cent. It was almost as large in the north-western region, but in the central black earth region there was little change until 1958 and a substantial contraction afterwards.

The geographical dynamics of the collective farm conversion programme may, therefore, be summed up as follows. It started in 1956-57, when conversions took place in all parts of the Russian Federated Republic, in the north-west and, perhaps, in Uzbekistan. During the four years 1958-62, conversions reached mass proportions in Russia, except such traditional strongholds of grain production as the lower Volga and north Caucasus regions but including the central black earth zone; very sharp reductions in the number of collective farms also took place in the north-west, White Russia and the Baltic republics, while in Kazakhstan the number of collective farm households was reduced almost as severely as in Siberia. On the other hand, the European South (Ukraine and Moldavia), the Transcaucasus and central Asia remained virtually unaffected by this policy. During the last two years of the Khrushchev era, reductions in the number of collective farm households continued on a very much reduced scale and in an irregular fashion. Thus there were quite significant falls in two of the Transcaucasian republics which previously had remained almost immune to the process, while the opposite was the case in central Asia.

The number of collective farm households is a reasonably objective measure of changes in geographical location, although in a rapidly expanding industrial economy some reduction in the absolute number of peasant households is only to be expected and this may vary in different areas. The financial status of the converted collective farms may be roughly expressed in terms of their money income per peasant household; this can be done on the rough-and-ready assumption that the net decline in the number of collective farm households was, on balance, entirely due to the conversion of collective farms into State farms.

The number of collective farm households in 1959 was 18.5m., and 1.2m. had been affected by conversion into State farms. The collective farms still in existence in 1959 had a money income of 4,400m. (new) rubles in 1953, which was equivalent to 238 per collective farm household. As the total money income of *all* collective farms in 1953 had been 4,960m. rubles, those converted into State farms six years later had a total money income of 560m. in 1953, equivalent to 467 rubles per household.[26] However crude the method of calculation, it is clear that the collective farms converted into State farms during the early years of the conversion policy were, on balance, considerably better off than the average.

Similarly, it may be estimated that the conversions during 1959 to 1962 affected about 2.2m. households[27] and that the average 1953 money income of the collective farms affected by this process was 250 rubles per household, compared with 237 rubles per household in 1953 for the 16.3m. households still belonging to collective farms in 1962.[28] Thus there was no significant difference in the prosperity, as measured by money incomes per household, between the collective farms converted into State farms during the peak period of this policy and the rest.

For the 400,000 collective farm households affected, on balance, by conversion between 1962 and 1964, such figures cannot be calculated for 1953 but only for 1958. In that year the money income of the collective farms still in existence in 1962 was 11,400m. rubles and that of the collective farms remaining in 1964 was 11,200m. rubles, or 700 rubles per household, but that of the 'converted' collective farm households only 500 rubles.[29] Thus at a point in time well before their conversion into State farms the collective farms converted between 1962 and 1964 were substantially less prosperous than the average.

The results obtained by this rather tedious but nevertheless very summary method are surprisingly clear-cut. The conversion policy started with comparatively wealthy collective farms, its main weight affected collective farms of average money incomes —which suggests that at that time the policy was not related to the prosperity or poverty of the collective farms concerned— and its final stage concentrated on units which were considerably poorer than the average.

The execution of the conversion policy was accompanied by official assurances that the Government had no intention to abolish 'collective-co-operative property' but that the aim was some form of fusion between all forms of public property.[30] There certainly is no good reason to regard this policy as ideologically inspired: otherwise no part of the Soviet Union would have escaped it, for politically motivated campaigns are necessarily universal. The claim that the conversion of collective farms into State farms was an 'enserfment of the rural population'[31] would have made sense only on the unrealistic assumption that the collective farms were in fact as well as in name producer co-operatives. This would be no more justified than to assume that the normal formula used in collective farm legislation of 'advice' to the general members' meeting expressed genuine deference to the wishes of the peasants.

A rational interpretation of the conversion policy must be based on the fact that the post-Stalin era witnessed a fairly radical change in the financial and economic relations between the Soviet power and both forms of public agriculture. The policy of crediting the State farms for their deliveries of agricultural produce at almost nominal prices and meeting their inevitable operating losses from the budget was changed in April 1954, when it was decided to allow them to run at a reasonable profit.[32] (The decree on the development of the State farms issued on April 15, 1954, replaced Government financing by new acquisition prices from the beginning of the year.) At that time the first steps had already been taken towards raising Government procurement prices and this policy was steadily extended during the next few years.

Under Stalin, the peasants had to accept the use of the collective farms as a means of exploitation by the Government which extracted a large part of the product of their labour without economic equivalent. Their co-operation in this process was, to say the least, unenthusiastic and their grudgingly performed labour was lavishly extravagant in relation to the results achieved. During the post-war years, the cumulative effects of war-time population losses and the incipient repercussions of the low war-time birth rate on labour supplies became an increasingly important economic issue.[33] In such conditions, the artificial cheapness of the collective farms as a source of supplies—which was in any case losing some of its attractions with the steady rise in producer prices—became less important than the fact that comparisons of labour usage in physical terms tended to favour the State farms *vis-à-vis* the collective farms.

TABLE 40
Direct labour expenditure for selected products, 1956–57
(hours per quintal)

	State farms	Collective farms
Grain	1.8	7.3
Potatoes	4.2	5.1
Sugar-beet	2.1	3.1
Raw cotton	29.8	42.8
Cattle*	52	112
Milk	9.9	14.7
Pigs*	43	103

* Weight gain.
Source: Selkhoz, p. 449.

Such average figures may be subject to severe qualifications, but they must have suggested to the Government two things. The first was that State farms managed to produce the main agricultural commodities with a considerably lower expenditure of labour than collective farms; the second was that this advantage was particularly pronounced in the case of grain production.

There were no doubt objective reasons for the particularly poor performance of collective farm labour, such as the lower degree of mechanization of the collective farms. However, there was also the crucial fact that the collective farm worker was expected to work for many months before he received his labour day 'dividend' from the collective farm. A change in the system of payment, involving regular monthly advances on his final entitlement, could be expected to give the collective farmer a specific incentive for working more efficiently. The collective farms were, therefore, 'encouraged' to make such advance payments and inevitably many of them found it impossible to combine adequate incentives to their members with financial solvency.

If this half-way house between the original concept of the collective farm and the treatment of the peasant as a worker, entitled to proper wages for proper work, proved uninhabitable, conversion of the collective farm into a State farm must have appeared as the obvious way of raising the productivity of agricultural labour and freeing under-employed agricultural labour for industry, while at the same time expanding agricultural production. The number of persons employed in public agriculture rose only from 26.2m. in 1953 to 28.2m. in 1956. From then until 1958 there was a fall in the agricultural labour force to 27.3m., due to a fall of 2.5m. in collective farms (and MTS) which was only partly compensated by increases elsewhere. During the next four years this process gathered momentum: collective farm labour contracted by about 5m., while State farm staffs rose by 2.8m., a net decline of over 2m. During 1962-64, on the other hand, a further fall of 400,000 in the collective farms was exactly compensated by a rise in the State farms. The drastic fall in the agricultural labour force was thus over, at least for the time being.[34]

The gradual abandonment of the conversion policy after 1962 may, therefore, have been connected with the fact that the anticipated labour shortage in industry through the lower war-time birth rate did not, in fact, materialize.[35] Nevertheless the better use of the increasingly valuable man (and woman) power must be rated very high amongst the economic motives of the conversion policy. In narrower terms, the Government was also influenced by the fact that during the 1950s the cost of procuring agricultural produce from the State farms was much lower than from the collective farms. Khrushchev claimed as late as 1957 that the Government paid the State farms 172 rubles per 100 kg. of raw cotton compared with 352 rubles for the collective farms; for sugar-beet the comparable prices were 14.4 and 26.2 rubles and for sunflower seed 52 and 146 rubles.[36] Though these figures must not be taken as a simple reflection of State farm and collective farm costs, it is probable that with the absorption of the less efficient collective farms by the State farms the cost of production in the latter tended to go up much more than in the remaining collective farms. Thus the average cost of cotton production in the State farms rose from 209.3 (old) rubles in 1959 to 284.8 in 1962 compared with an increase from 200.3 to 223.9 rubles in the collective farms.[37]

There were, however, some specific factors which played a considerable part in the decision to push forward with the formation of State farms, including the conversion of existing collective farms. With a backward and over-extended transport system, the proximity value of perishable produce, such as fruit and vegetables, meat and milk, for the supply of the towns was high and reliance on the produce of the tiny subsidiary plots of the collective farmers, or even on the surplus of the collective farms, was risky and expensive. Hence the Government decided to set up a ring of State farms around the largest towns. In 1956, the number of State farms specializing on meat and milk was actually lower than it had been in 1950, but in the next two years it rose by over 30 per cent; though the increase in the number of specialized fruit State farms began later, it amounted to almost 50 per cent in the single year 1959.

TABLE 41

Number of State farms by main enterprise, 1953–64

Main enterprise	1953	1956	1958	1962	1964
Grain	477	844	1,036	1,135	1,229
Sugar-beet	217	206	208	277	316
Cotton, etc.	125	133	198	214	247
Fruit and vegetables	806	784	816	1,222	1,548
Milk and Meat	1,284	1,321	1,739	3,170	3,735
Pigs	718	619	622	573	581
Poultry	177	167	202	443	613
Sheep	341	435	610	772	987
Other livestock	240	158	168	234	228
Total (including others)	4,857	5,098	6,002	8,570	10,078

Sources: Selkhoz, p. 42, Narkhoz 1964, p. 408.

Up to 1956, before the start of the conversion policy, the increase in the number of State farms was concentrated on grain production—above all in Kazakhstan, where it more than doubled.[38] Between 1956 and 1958, grain farms certainly continued to increase, cotton and sheep farms increased even faster and there was a spectacular jump in the number of cattle farms (meat and milk). From then onwards, fruit and vegetable farms and sugar-beet farms made the running amongst arable enterprises and dairy farms as well as poultry farms among livestock enterprises. At the same time there was a marginal withdrawal of State farms from pig-keeping and, inevitably, from horse-breeding.

The production strategy and the organizational policy of the Khrushchev era went, broadly, hand-in-hand. A growing proportion of food grain production and, above all, of wheat was concentrated in the 'new lands' in the East and South-east, while the traditional grain areas were directed more towards the production of fodder and livestock enterprises. As part of this process, food grain production became increasingly the business of the State farms, while the collective farms were permitted, and indeed 'encouraged', to use a greater part of their diminished resources for feed grain and fodder crops. With the conversion of collective farms after 1956, State farms began to take a quickly increasing part in livestock production, with cattle, sheep and poultry farms all expanding from year to year in order to provide the authorities with greater supplies of livestock products from enterprises under their direct and complete control.

3. PRODUCTION AND MARKET SUPPLIES

Two Steps Forward, One Back

The identification of the top leadership with a dynamic agricultural policy involving a vigorous rise in capital investment and substantially higher producer prices, coupled with the initial success of the 'new lands' campaign, resulted in a considerable rise in output between 1953 and 1958. As 1958 was climatically exceptionally favourable, it is fairer to compare average crop figures for 1949-53 with the average for 1954-58, but the evidence remains impressive, whatever the basis of comparison. (Appendix, Tables II and III.)

THE KHRUSHCHEV ERA (1953-1964)

With the exception of feed grains, other than maize, potatoes and cotton, gross production of the main commodity categories generally increased by over a third, and in numerous cases considerably more. The main reservation concerns the official grain statistics which were claimed at the time to be 'barn weights' but were widely suspected by non-Soviet experts and revealed as 'bunker weights' by the 1964 official statistics—a difference which has been estimated to amount to as much as 20 per cent.[39]

However, at the time, such exaggerations were at the most suspected and the successes were plain for all to see. No wonder that the leading actor in this transformation scene was able to sweep aside all domestic opposition to policies which paid such handsome dividends and to proclaim the imminence of the moment when the Soviet Union would 'reach and surpass' the United States output of meat and butter.

Unfortunately it soon became clear that 1958 had produced a record harvest but no break-through in the sense of self-sustaining growth in crop production and, above all, in cereals. It was much more like a traditional good season, owed to the fortuitous bounty of a generally niggardly climate and not to policies which could produce acceptable results even in unfavourable climatic conditions. The crest of the wave was followed by a trough and by a series of fluctuating crop yields on a moderately lower level, until the catastrophic season of 1963 seemed to put in question all that had been achieved before. Though it was followed by another bumper season, its crop was already garnered by Khrushchev's successors.

TABLE 42
Main crop yields, 1953–63
(quintals per hectare)

	1953	1958	1962	1963
All cereals	7.8	11.1	10.9	8.2
incl. winter wheat	11.1	16.2	16.8	12.9
spring wheat	7.0	9.7	8.2	5.9
maize (full grain)	10.6	23.3	22.1	15.9
spring barley	8.1	13.1	11.6	9.0
Cotton	20.5	20.2	18.0	21.0
Sugar-beet	148	218	152	120
Sunflower seed	6.7	11.8	10.9	9.7
Potatoes	87	91	80	84

Source: Selkhoz, p. 209. Narkhoz 1964, pp. 295ff.

Throughout the second half of the Khrushchev era, gross production of wheat never again reached the 1958 level, while the only other cereal to make substantial, and indeed excessive, progress was maize. Raw cotton production stagnated until 1963, sugar-beet crops first stagnated and then fell sharply, even in advance of the crisis year 1963.

The poor performance of the years 1959 to 1963 was essentially due to the failure to repeat—not to mention to improve upon—the relatively high yields obtained in 1958, though some part of the advance made since 1953 was generally retained.

The output of livestock products depends in the Soviet Union more on the previous year's harvest than in countries blessed with a more equable climate. It benefited in 1959 from the large 1958 crop, declined marginally in 1960 after the poor 1959 harvest, rose more than proportionately to the rise in fodder production during the next two years and suffered a severe, though temporary, setback in 1963 and 1964. On balance, meat production increased quite substantially over the 1958 level, even if the involuntary and temporary rise in 1963 is discounted, because it was the result of the emergency slaughter of pigs for which not enough feed was available. Milk production, on the other hand, stagnated from 1959 onwards, wool production did not fare any better, and only the output of eggs rose steadily until 1962, though feed difficulties during the next two years caused an appreciable drop in supplies.

At least until 1963, livestock population increased much more quickly than the output of livestock products. (Appendix, Tables III and IV.) Even after the enforced cut in livestock numbers after the poor 1963 harvest, the national herd of cows was 15 per cent, and that of other cattle 26 per cent, greater than in 1958, but of smaller livestock only sheep retained some of the gains made after 1958, while pigs and poultry numbers were drastically reduced.

This dramatic emergency was a painful demonstration of the fact that during the preceding period livestock numbers had expanded beyond the capacity of the supporting feed base.[40] In terms of feed units, public agriculture disposed of 167m. tons in 1964 compared with 190m. tons in 1962.[41] Although silage-making as a relatively new technique could have been expected

190

to make quick progress, only 137m. tons was produced in State and collective farms in 1964 compared with 148m. tons in 1958.[42] The inevitable result was a drop in the yield of animal products: the average milk yield per cow for the whole of agriculture in 1964 was noticeably lower than in 1958 (1,655 kg. compared with 1,755 kg.), and the fall was particularly steep in collective farms. Similarly, the fleece weight per sheep dropped from 2.7 kg. in 1958 to 2.5 kg. in 1964, when it was no higher than it had been in 1956.

Market Supplies and Food Consumption

The trend of agricultural gross production was, in the last resort, the main determinant of the volume of market supplies for the urban population and for industry. With a gently declining rural population and a substantial rise in gross production up to 1958 a sharp rise in market supplies could be expected, and this is what happened. The urban population rose between 1953 and 1958 from 80.2m. to 95.6m. and continued to rise to 118.5m. in 1964—a fairly steady increase of 3.7 per cent per year. In order to provide rising supplies per head of the urban popula-

TABLE 43
Market supplies in 1953, 1958 and 1964

	(i) in m. tons (except eggs and wool)			(ii) % change per year	
				1958 over	1964 over
	1953	1958	1964	1953	1958
Wheat	21.8	46.4	40.7	16.3	−2.2
Other grain	14.0	16.0	33.4	2.7	13.0
Total grain	35.8	62.4	74.1	11.8	2.9
Raw cotton	3.85	4.34	5.28	2.4	3.2
Sugar-beet	22.9	51.0	76.1	17.4	6.9
Sunflower seed	2.07	2.98	4.14	7.5	5.6
Potatoes	12.1	14.1	16.6	3.1	2.8
Vegetables	5.1	7.1	10.4	6.8	6.6
Meat	3.2	4.8	5.8	8.5	3.2
Milk	13.7	25.4	34.2	13.1	5.1
Eggs*	5.8	8.0	11.3	6.7	5.9
Wool†	198	278	310	7.0	1.8

* '000m.   † '000 tons.
Sources: Narkhoz 1962, p. 233. Narkhoz 1964, p. 253.

tion, market supplies, therefore, had to rise by a higher percentage, though this oversimplifies the situation somewhat.

To compare individual points in time can only give a very approximate picture; as 1953 was a very mediocre year in every respect and 1958 very favourable, such simple comparisons may give an excessively rosy picture of the progress made in the first five years of Khrushchev's ascendancy and a slightly pessimistic assessment of the later period. The use of average annual rates of change also suggests a steadier pace of improvement in supplies than was achieved in practice. Despite such reservations it remains true that the period as a whole witnessed sustained, if uneven, progress in the supply of basic agricultural raw materials as well as the main foodstuffs. For grain, sugar-beet, sunflower seed, potatoes and livestock products the rate of increase slackened appreciably after 1958; on the other hand, the supply of raw cotton increased marginally more quickly in the second half of the Khrushchev era than in the first. (Comparisons between 1958 and 1963 would show slightly more favourable results for livestock products, except milk, and drastically worse figures for most crops.)

There were, however, some foodstuffs for which the supply per head of the town population fell—potatoes throughout the whole era, grain (and particularly wheat) and meat after 1958. By far the most important of these was grain, though in this case the significance of the measure adopted here is not particularly great. However it is a fact that in 1958 the Soviet Union was a net exporter of grain to the extent of 4.3m. tons; grain exports rose further to 6.6m. tons in 1960 and reached a peak in 1962, but in 1963 gross exports (largely to other Soviet bloc countries) amounting to 6.3m. tons were accompanied by imports of 3.1m. tons from the West (Canada, the United States, Australia, Argentina, etc), while in 1964 the Soviet Union was actually a net grain importer for 3.8m. tons.[43]

This spectacular, and for Khrushchev disastrous, reversal of Russia's traditional position in the international grain trade should not be allowed to overshadow the remarkable progress made since Stalin's death. Food supplies in the Soviet Union during this period were neither inadequate nor even unsatisfactory except against the background of rapidly increasing incomes and a correspondingly voracious demand for high-quality foodstuffs. The estimate of food supplies provided by the United States Department of Agriculture, though described as 'experimental and tentative', makes it clear that the Soviet Union was amongst the best-fed countries in the world, though this was less true in relation to quality than to quantity.

TABLE 44

Net food supplies per head of the population, 1959–61

	(per year) Product weight (kg.)	(per day) Calories	(per day) Protein (gr.)	(per day) Fat (gr.)
Total cereals	171.4	1,631	45.2	6.7
Sugar, raw	30.1	319	—	—
Potatoes	138.0	265	6.4	0.4
Vegetables	64.8	78	5.0	0.6
Fruit	17.0	22	0.3	0.1
Beef and veal	12.2	53	5.1	3.7
Pork	12.4	102	3.7	9.2
Other meat	12.9	46	5.0	2.7
Fish	11.4	19	2.7	0.8
Eggs	6.7	26	2.0	1.9
Butter*	3.8	74	0.1	8.4
Other fats*	7.5	176	0.2	19.7
Whole milk	101.2	180	9.7	9.7
Cheese	0.9	7	0.5	0.6
Total consumption		ca. 3,000	85.9	64.5

* Fat content.

Source: U.S.D.A.—Food Balances in Eastern European Countries (1965).

Despite the disproportionately high share of cereals and potatoes in the supply of calories and the relatively low proportion of animal protein in the diet, the feeding standard of the Soviet population firmly places it amongst the minority with ample calories and adequate protein resources at its disposal. Not even the poor harvest of 1963 caused a substantial fall in food consumption per head, because the Soviet Government could afford to make good the deficiency in wheat supplies through imports. Food consumption per head in 1963-64 has been estimated at 2,950 calories per day, with 90 grammes of protein and 75 grammes of fat.[44]

Official Soviet statistics are available only in index form and such figures defy detailed interpretation, but they suggest a very large rise in living standards between 1940 and 1960 and an appreciable rise between 1953 and 1960, with a shift from starch to sugar as a source of carbohydrates, a large relative improvement in the consumption of high-protein foods and a very substantial expansion of mass consumer goods, presumably from a very low base level.

TABLE 45

Personal consumption of important consumer goods (1940 = 100)
(based on family budgets)

	Workers' families		Peasants' families	
	1953	1960	1953	1960
Bread and flour products	87	74	98	93
Meat (incl. poultry)	166	247	110	186
Fish and fish products	140	174	155	323
Milk and milk products*	170	265	110	161
Eggs	148	244	140	288
Sugar	196	225	288	739
Confectionery	140	188	170	323
Textile materials	207	162	171	194
Clothing	222	325	103	272
Shoes	148	160	134	191
Furniture and household goods	276	591	157	408
Cultural goods	596	1,654	247	1,334

* Milk equivalent.
Source: Narkhoz 1960, pp. 742f.

4. THE ALLOCATION OF RESOURCES TO AGRICULTURE

Before collectivization, the Government could obtain agricultural produce from the peasants either through taxation or through exchange. Direct taxation involved severe administrative difficulties, as taxes had to be collected from twenty-five million households; indirect taxation, though very simple to apply, was subject to the law of diminishing returns. When prices of industrial goods were set too high or those of agricultural products depressed too much, the régime ran the risk of a peasants' strike as buyers and sellers which endangered the food supply of the towns.

The peasants, therefore, enjoyed a high degree of autonomy

in dividing their final output between consumption in kind, accumulation (e.g, in the form of livestock) and market produce. Within the limits of availability of suitable supplies they could also determine freely how much produce they wanted to exchange for producer goods, either for current inputs or for investment, and how much for consumer goods.

It was the main, if unavowed, purpose of collectivization to remove this discretionary power from the peasants and to make the delivery of planned, i.e, predetermined, quantities of 'marketable surplus' a prior charge on the agricultural economy, irrespective of the volume of industrial goods available in exchange. The system painfully and wastefully evolved after 1928 transferred part of the cost of capital investment (in tractors and certain other forms of agricultural machinery) to the Government and put most of the risks from fluctuations in farm output, as well as a heavy tax burden, on the peasants. The impasse reached at the time of Stalin's death strongly indicated that the volume of investment was too low to permit expansion of production and that the income accruing to the peasants from their work in the collective farms was insufficient to provide incentives for better work. The conclusion was inevitable that more resources had to be made available in the form of more capital investment and higher producer prices.

Capital Investment in Agriculture

Within certain limits, global figures provide a useful guide to the place accorded to agriculture in the hierarchy of economic aims and a convenient overall measure of its relative importance. Table 46 overleaf shows the distribution of capital investment from all sources, including the peasants in their individual and collective capacities, by main sectors.

The sharp decline in the share of agriculture between 1927-28 and 1950 was, of course, implicit in the industrialization process, though its precipitous drop may reflect the fact that agriculture was being starved of essential resources during most of this period. The very high level of 'other' investment (mainly in houses and other buildings) during the 1950s reflects mainly the enormous reconstruction tasks after the Second World War. (The uncertainty about the size and treatment of the very large

TABLE 46

The structure of gross capital investment
(in % of total)

	Industry	Agriculture	Transport	Other	Total
1927–28	27.7	43.4	12.7	16.2	100.0
1950	40.8	15.0	12.2	32.0	100.0
1952	45.5	13.7	9.1	31.7	100.0
1958	37.4	15.8	7.5	39.3	100.0
1959	38.2	14.9	8.0	38.9	100.0
1960	39.0	14.2	8.5	38.3	100.0
1961	39.0	15.0	8.6	37.4	100.0
1962	39.3	16.0	9.0	35.7	100.0
1963	38.4	16.5	10.0	35.1	100.0
1964	39.5	18.0	9.9	32.6	100.0

Sources: 1927-28: *Piatiletni Plan* I, 139. 1952-62: Narkhoz 1962, p. 434.
1950, 1963-4: Narkhoz 1964, p. 514.

TABLE 47

Productive capital investment in agriculture
(comparable prices, '000 m. 'new' rubles)

	Total	Public funds	Collective farm funds	Collective farm funds in % of total
Average 1929–32*	0.28	0.21	0.07	25
Average 1933–37	0.44	0.24	0.20	45
Average 1938–41†	0.60	0.23	0.37	62
Average 1941–45‡	0.40	0.07	0.33	82
Average 1946–50	1.12	0.50	0.62	55
1951	1.94	1.02	0.92	47
1952	2.04	0.97	1.07	52
1953	2.06	0.88	1.18	57
1954	2.97	1.54	1.43	48
1955	4.11	1.99	2.12	51
1956	4.38	2.12	2.26	52
1957	4.54	2.34	2.20	48
1958	5.12	2.28	2.84	56
1959	5.55	2.02	3.53	64
1960	5.64	2.47	3.17	56
1961	6.14	2.98	3.16	51
1962	6.66	3.39	3.27	49
1963	7.32	3.90	3.42	47
1964	8.73	4.82	3.91	45

*Including last quarter of 1928. † Up to the outbreak of war.
‡ War period, July 1941 to December 1945.
Sources: Up to 1950: Selkhoz, p. 387. From 1951: Narkhoz 1964, p. 517.

resources devoted to all branches of military expenditure, including the atomic weapon system and the space programme, inevitably limits conclusions to very broad statements.) Nevertheless, during the early years of the Khrushchev era the share of agriculture rose while that of other branches of the economy, such as industry and transport, sharply declined. However, there was a noticeable drop in 1959 and 1960, and even in 1961 the proportion of agricultural investment remained well below the 1958 level; from 1962, on the other hand, there was a substantial and sustained increase in the share of agricultural investment.

Within the total of agricultural capital investment there were quite substantial shifts between the amounts raised by the Government and those contributed by the collective farms, i.e, in the last resort by the peasants themselves.

There was a steep rise in investment, both from public sources and from collective farm funds, after 1953 which lasted until 1957. In 1958 and 1959 public investments declined somewhat, while investments by the collective farms increased enormously and reached a peak in 1959. This increase is even more remarkable, because it excludes the 1,800m. rubles paid by the collective farms for the assets of the Machine and Tractor Stations which they acquired during 1958 and 1959. From 1960 onwards, public investments began to rise again, while collective farm investments fluctuated around a somewhat lower level and did not exceed their 1959 figure until 1964.

A substantial part of productive investments took the form of agricultural plant and machinery, and the value of the output of agricultural machinery as a whole fell from 948m. rubles in 1957 to 674m. in 1959 and barely exceeded the 1957 level as late as 1961, when it reached 972m. rubles.[45] Given the critical importance of these items in large-scale Soviet agriculture, an attempt to trace changes in the numbers of individual types actually available during this period may be worthwhile, even though it is limited to the years from 1957 onwards.

New supplies of tractors stagnated from 1957 to 1960, those of grain combines declined sharply from 1957 to 1958 and remained at the lower level until 1960, those of lorries and tractor-cultivators fell progressively from 1957 to 1960, and only new supplies of tractor-ploughs rose from 1957 to 1958 before de-

TABLE 48

Supplies of selected types of agricultural plant ('000 units)

	Tractors	Tractor-ploughs	Tractor-cultivators	Grain combines	Lorries
		(i) New deliveries to agriculture			
1957	148.3	128.5	207.5	133.7	125.3
1958	157.5	160.3	164.2	64.9	102.1
1959	144.3	145.1	123.2	53.1	76.3
1960	157.0	142.4	79.2	57.0	66.1
1961	185.3	133.1	99.4	70.0	69.7
1962	206.0	133.7	118.7	79.2	82.6
1963	239.3	172.1	153.7	79.6	68.8
1964	222.5	174.6	186.4	78.6	63.0
		(ii) Net increase in plant resources			
1957	54.0	−7.0	116.0	108.0	29.0
1958	77.4	57.0	28.0	18.7	40.0
1959	52.6	{ 14.0	{ −112.0	−17.7	29.0
1960	68.3			3.2	49.0
1961	89.7	8.0	4.0	0.8	18.0
1962	116.9	40.0	58.0	21.6	79.0
1963	113.1	54.0	−3.0	−2.4	47.0
1964	97.0	23.0	50.0	−4.6	32.0

Sources: Selkhoz, pp. 415, 419. Narkhoz 1962, pp. 326, 329. Narkhoz 1964, pp. 384, 389 and calculations therefrom.

clining during the succeeding years. After making allowance for wastage through obsolescence and wear and tear, the number of tractors rose modestly from year to year until 1960 and rather more quickly from then onwards, and the number of lorries also increased, though in some years by very little. On the other hand, the stock of grain combines actually fell after 1958 and regained the earlier level only in 1962, that of tractor-cultivators dropped sharply between 1958 and 1960 and that of tractor-ploughs changed very little between 1958 and 1961.

There was thus a very noticeable slow-down in the mechanization of agriculture during the years 1958 to 1961. This resulted from a fall in the delivery rate of many important items and from an increase in wastage. The former was connected with important developments outside the sphere of agriculture itself—the competing requirements of the armaments and space race with the United States, on the one hand, and the pressure for greater supplies of consumer durables to the more prosperous sections of the town population on the other. A spectacular instance of the latter was the switch from the manufacture of combine harvesters by the agricultural plant at Zaporozhe to mini-cars (the Zaporozhets) which aroused Khrushchev's anger in 1962[46]—and illustrated the limitations of his power within the Soviet leadership.

The increased wastage rate which can be deduced from the official data for 1958 and the immediately succeeding years may well have been connected with the transfer of the MTS plant to the collective farms. An occasion of this kind is bound to lead to more thorough stock-taking, probably involving the writing off of a substantial part of the inventory, either because it was no longer in existence or because its technical usefulness was at an end.

Whatever the real cause, the net result was a formidable degree of under-equipment of public agriculture with machinery of all kinds. It was officially estimated in 1962 that the stock of tractors at the beginning of the year, at 1.17m., compared with a requirement of 2.7m., that there were 845,000 grain combines needed but only 503,000 available, that agriculture required 1,650,000 lorries but only had 790,000, while the shortfall in most of the more sophisticated items of machinery was even greater.[47]

The relaxation of efforts after the great crop year of 1958 was not confined to agricultural plant. Between 1953 and 1958 deliveries of mineral fertilizers to agriculture increased from 6.6m. tons to 10.6m., or by an annual average of 800,000 tons per year; this was equivalent to a growth rate of about 10 per cent compound per year. The increase in 1959 was less than 500,000 tons (5 per cent) and this was followed by an increase of less than 300,000 tons (under 3 per cent) in 1960, before rising to 700,000 tons (6 per cent) in 1961.[48]

In most areas where it is possible to amplify global figures by more detailed analysis, a similar picture emerges. There was a great spurt in the resources devoted to agriculture, and a corresponding rise in activity, during 1954 to 1957 or 1958, followed by a slackening of the rate of increase during the following two or three years. In the event, the régime was faced with the same disheartening spectacle of unsatisfied needs in 1962 which had confronted it on the morrow of Stalin's departure from the scene, although in absolute terms very substantial progress had been made.

The Restoration of Price Incentives

The reinstatement of producer prices as an effective part of agricultural policy was one of the earliest and most sustained changes introduced by the new leadership. Stalin's last publication and a series of later revelations confirm the serious disagreement between the dictator and the reformers on this subject during the early 1950s, and Khrushchev's keynote speech of September 1953 announced substantial price increases, particularly for livestock products.

Under Stalin there had been a strong price emphasis on certain industrial crops, particularly cotton but also tobacco and sugar-beet. During the years 1953-58 prices for these products rose very little and the increases benefited mainly grain, sunflower seeds, potatoes and meat. In addition, the distinction between compulsory quotas at almost nominal prices and excess deliveries at more remunerative prices was increasingly seen to operate not only inequitably but also inefficiently and was abolished in 1958, though this proved to be only temporary. The main arguments against the dual pricing system were twofold: it benefited collective farms on rich soils and discriminated against those on poorer land, and it had the paradoxical effect of producing higher average producer returns per unit in good years than in bad seasons.

The extent of the price increases carried out under Khrushchev can be seen by comparing the prices paid by public trade in the last year of the Stalin era and in 1963. Though the absolute figures for 1958 are not strictly comparable with those for 1952 and 1963, because they are average prices obtained for agricultural produce as a whole, they are included as a matter of interest; price indices refer throughout to procurement prices of public trade.

The process of price adjustment was virtually complete by 1958 for cereals (excluding maize, which was given preferential treatment) and sunflower seed; it was, indeed, partly reversed during the following few years for wheat and sugar-beet. Other important products for which price increases between 1958 and

TABLE 49

Agricultural producer prices, 1952–63

	(i) in new rubles p. 100 kg.			(ii) 1952=100			(iii) % rise p.a.	
	1953	1958	1963	1953	1958	1963	1958/ 1952	1963/ 1958
Wheat	0.97	—	7.56	245	621	779	35.6	4.6
Maize (grain)	0.54	—	7.66	207	819	1,419	42.0	11.6
All grain	—	5.80	—	236	695	—	—	—
Sugar-beet	1.05	2.25	2.87	144	219	273	14.0	4.5
Raw cotton	31.88	31.7	38.30	105	106	120	1.0	2.5
Sunflower seed	1.92	16.77	18.10	528	774	943	40.6	4.0
Tobacco	72.03	115.88	176.65	96	162	245	8.4	8.6
Potatoes	0.47	8.0	7.10	316	789	1,511	41.1	16.4
Vegetables	1.92	11.42	7.52	—	—	392	—	—
Cattle	2.03	54.95	79.90	338	1,147	3,936	50.2	27.9
Pigs	6.72	88.59	98.00	453	1,156	1,458	50.4	4.8
Milk	2.52	12.47	12.18	202	404	483	26.2	3.6
Eggs*	19.90	77.70	70.00	126	297	352	19.9	3.5
Raw wool	106.80	328.00	378.67	107	352	355	23.3	0.2

* 000 units.

Sources: The absolute figures for 1952 and 1963 are the procurement prices paid to collective farms and collective farmers, etc., per Khrushchev *op. cit.*, (in III, II), Vol. VIII, 467. The 1958 figures are the average prices for agriculture as a whole per *Ekonomicheskoie obosnovanie struktury selskovo khoziaistva* (1965), p. 170. The index figures for 1953 and 1958 from Selkhoz, p. 117, for 1963 calculated from the absolute figures for 1952 and 1963; the annual compound increases (per cent) calculated from the index figures.

1963 were modest or insignificant were wool and cotton, milk and eggs, pigs and sugar-beet. On the other hand, tobacco, potatoes and, above all, cattle continued to receive substantial price boosts, though with the exception of tobacco the rate of improvement was considerably smaller than during the previous five years.

The abysmally low acquisition prices paid under Stalin for all except a few favoured commodities permitted (and required) phenomenal rates of increase during the years up to 1958, when State procurement prices were on average for all products about three times the 1952 level, with average annual rates of increase of 40 or 50 per cent for a number of products. During the next five years there were some price increases—as well as decreases —but these were generally only a fraction of those accorded during the preceding five years.

The higher procurement prices applied, of course, to the produce of the individual plots of the peasants as well as to the collective farms and made disposal through the channel of public trade financially more attractive to both types of suppliers. At the same time, the improvement in the supply of agricultural products from 1953 to 1962 was reflected in the price trends on the collective farm market. The prices of staple foods, such as bread, vegetable oil and butter were, in fact, lower in 1962 than ten years earlier, those of eggs, potatoes and vegetables were only moderately higher and the only really worthwhile increases were obtained for fruit, beef and poultry.[49] The harvest failure of 1963 and the resulting shortage of livestock products lifted collective farm market prices in 1964 to scarcity levels,[50] but this temporary change does not seriously affect the overall conclusion that the collective farm market had become a considerably less important source of income for the farm population than it had been in the early post-war era.

The increased producer prices paid by the State and the stagnation in receipts from the collective farm market were particularly important for the collective farms; unlike the State farms (and the MTS until their disbandment in 1958), they were expected to balance their accounts by varying their disbursements for labour days with the excess of their receipts over their expenses and the income of the collective farm members from their work for the collective farms depended, therefore, mainly on the cash income of the latter, though they also received certain payments in kind.

The money economy of the collective farms was massively transformed in the five years after Stalin's death. Their total cash income rose threefold and the bulk of the increase consisted in higher returns from public trade, with receipts from livestock products expanding much faster than those from crops. After 1958—which benefited considerably from an excellent harvest—cash receipts stagnated for three years in precisely the same way as agricultural gross production and investment. The first major re-expansion of cash income took place in 1962, largely due to the increased meat and butter prices in June 1962. Receipts from the collective farm market, on the other hand, were only fractionally higher in 1958 than they had been in 1952 and fell appreciably until 1962, before regaining the 1958 level in 1964.

TABLE 50

Agricultural money income of the collective farms, 1952–64
(i) in '000 m. new rubles

	From all sources			From public trade	From collective farm market
	Crops	Livestock products	Total		
1952	2.9	1.0	3.9	2.4	1.5
1953	2.9	1.7	4.6	3.2	1.4
1958	7.6	4.8	12.4	10.8	1.6
1960	7.2	5.3	12.5	10.9	1.6
1962	8.2	6.2	14.4	13.1	1.3
1964	10.9	6.0	16.9	15.3	1.6
		(ii) 1952=100			
1953	102	160	118	132	94
1958	267	451	318	443	107
1960	251	500	321	444	105
1962	286	590	369	536	84
1964	382	566	433	624	109

Sources: Up to 1962: Narkhoz 1962, p. 342. 1964: Narkhoz 1964, p. 400.
Note: Apart from the agricultural money income, collective farms earned substantial amounts from non-agricultural sources ranging from 400m. rubles in 1952 to 1,000m. rubles in 1964.

The official statistics of collective farm income represent gross receipts. However, farmers in Soviet Russia, as elsewhere, are in the last resort interested mainly in their margins, i.e. in the excess of gross receipts over disbursements. The prices which the collective farms had to pay for their inputs did not remain stable. The traditional discrimination against the collective farms in the prices charged for plant and machinery, as compared with the State farms, was abolished in 1958, and the prices of petrol and other mineral oil products were also put on the same basis. (In the first case all forms of public agriculture were now charged retail prices, in all other cases wholesale prices.) At the same time the prices of spare parts are said to have increased by 2 to 2.2 times.[51]

From the available evidence it is not possible to estimate the net effect of these cost changes on the net income of the collective farms, but it would be unwise to ignore the probability that

203

there was such a net effect; it is not unlikely that it was negative, for there is a fair amount of material suggesting that the sharp rise in farmers' incomes during the early years of the Khrushchev era, and particularly in the bumper year 1958, was regarded as excessive by an influential section of the ruling party. This attitude was also adopted by Khrushchev himself in his speech before the Central Committee in December 1959.[52]

Personal Earnings of the Farm Population
In 1953 the State farms, etc, employed 2.3m. people and the collective farms 23.9m. (including 1m. in the MTS) on agricultural work in the narrow sense of the term.[53] Ten years later, the Government-operated sector of State farms and similar enterprises had expanded to 7.3m. and the membership of the collective farms had declined to 17.6m. active workers.[54] Nevertheless, the collective farms continued to account for the large majority of agricultural income earners. State farm staff and collective farmers both obtained part of their income from public agriculture and part from their subsidiary private plots, but in other respects the differences between their income position were much greater than the similarities.

The staff of the State farms belongs to the working class (and the employed 'intelligentsia') and is in receipt of wages and normal social security benefits; the same was true of the staff of the MTS before their dissolution in 1958. Average earning figures are available only for selected years starting with 1958; they show monthly earnings both in rubles and in relation to those of industrial workers.

TABLE 51
Average monthly money earnings (in new rubles)

	Industry	State farms*	State farms in % of industry
1958	87.1	53.1	61
1960	91.3	53.9	59
1963	98.4	67.1	68
1964	100.5	70.6	70

* Including subsidiary State enterprises.
Source: Narkhoz 1964, p. 555.

Scanty though these figures are, they again convey the impression of absolute stagnation and relative decline after 1958 and of a sharp upward thrust in the closing years of the Khrushchev era. No detailed information is available about the earnings of State farm staff from their private plots, but Soviet sources state that earnings from the private economy amounted to 17 per cent of the final real income of the rural population as a whole and to considerably more for State farm workers and particularly for collective farmers.[55] If it is assumed for purposes of illustration that the earnings of State farm workers were 90 per cent of those of industrial workers and that the proportion of their income obtained from social security, etc, was about the same, this would imply that, perhaps, two-ninths of the earnings of State farm workers came from their subsidiary private economy and seven-ninths from their wages as State farm workers.

Though payment according to the number of 'labour days' continued to form the basis of the remuneration of collective farm members, the authorities began to experiment in the later 1950s with the payment of regular money advances on account of the final settlement. This move towards the payment of a regular wage to collective farmers was made possible by the substantial rise in producer prices and made necessary by the growing recognition of the fact that collective farm members were unwilling to exert themselves to the full in the course of the year for the sake of a low and uncertain reward at its end. The new system encountered difficulties, as it increased the financial risks for collective farm management and needed a better system of short-term credit than in the past; this was achieved to some extent by supplementing bank credit through advances from public purchasing organizations on account of future deliveries.

At least one of the reasons for the official reluctance to publish figures of average earnings per labour day was their extreme variability in different parts of the country. In 1953 Khrushchev indicated that there was an enormous range of payments according to farm type and location; while livestock farms paid 4 to 5 old rubles all told, sugar-beet farms in the Ukraine might pay 12 rubles, grain farms in the north Caucasus 18 rubles and cotton collective farms up to 36 rubles per labour day. Some years later he stated that the value of a labour day in the collective farms of the Moscow region had gone up from 1 ruble in 1953 to 5 rubles in 1957.[56] During the first years of the Khrushchev era, labour days became much more valuable with the rise in producer prices. In the Moscow region, the money payment in 1956 averaged 3.48 old rubles, or more than three and a half times the 1953 figure, the grain dividend at 1.16 kg. had risen similarly and the quantity of potatoes and vegetables showed on balance little change,[57] but the monetary value of payments in kind had risen with the rise in prices.

For comparisons between the earnings of collective farmers and State farm workers figures in terms of conventional labour days are useless and have to be replaced by figures of earnings per man day or per year. Yearly earnings depend to a large extent on the number of work days put in by the two groups of agricultural workers. As collective farmers worked on average only about two-thirds of the time put in by State farm workers in public agriculture,[58] their incomes from collective agriculture would have been considerably lower than those of State farm staff, even if they had earned the same amount per man day. In fact, this was by no means the case, though the difference was steadily shrinking. In 1962, the value of a man-day on collective farms was said to have been 40 per cent higher than in 1959 compared with an increase of 25 per cent on State farms;[59] this was claimed to mean that the 1962 value of a man-day on collective farms was 'near' that on State farms in 1958-60,[60] though this statement was, perhaps deliberately, vague.

On the other hand, it is known that in 1962 almost 44 per cent of all collective farms paid less than 50 per cent of the State farm wage to their members per man-day, a further 31 per cent paid between 50 and 75 per cent, 16 per cent between 75 and 100 per cent, and only 9 per cent paid more than the state farm wage.[61]

Regional differences (which reflected certain broad product specializations) always played an important part. Thus in 1956 and 1956-57 respectively, there were already some exceptions to the rule that daily earnings on collective farms were well below those on State farms. In Kazakhstan, State farms paid on average 22 old rubles per day, while collective farm members earned only 16.2 old rubles and State farm workers also fared better than collective farmers in the Krasnodar district (21.3 against 18.26 old rubles), Novosibirsk (18 against 15.4 old rubles) and Primorsk (22.2 against 18.75). On the other hand, in the Altai the collective farmers with 22.4 old rubles did much better than State farm workers with 18.4, while in Uzbekistan and the Stavropol district there was very little difference between them.[62] It will be seen that this process had progressed much further by 1964.

More satisfactory than the piecing together of different scraps of information, or of average figures whose validity cannot be properly assessed without missing background data, may be a comparison of production costs for individual commodities on collective farms calculated at State farm wage rates as well as at actual labour day payments.

TABLE 52

Estimated difference in earnings per man-day—State farms and collective farms (new rubles)

	1962			1964		
	Man days	Cost difference (r.)		Man days	Cost difference (r.)	
	per quintal	per quintal	per man-day	per quintal	per quintal	per day
Grain*	0.7	0.7	1.0	0.6	0.4	0.7
Cotton	6.9	2.3	0.3	6.0	—	—
Sugar-beet	0.4	0.3	0.7	0.3	0.1	0.3
Potatoes	0.9	1.0	1.1	0.6	0.5	0.8
Milk	2.7	2.7	1.0	2.7	1.1	0.4
Beef†	14.3	18.0	1.3	13.6	6.1	0.5
Pork†	14.5	20.1	1.4	13.8	7.6	0.5
Mutton†	10.2	5.9	0.6	6.6	2.8	0.4
Wool	47.2	24.4	0.5	48.1	11.1	0.2

* Excluding maize.   † Live weight.

Sources: Narkhoz 1962, pp. 338f, 362. Narkhoz 1964, pp. 394ff, 416.

Between 1962 and 1964 the differences between the earnings of State farm workers and collective farmers narrowed universally and disappeared for cotton. For grain, the difference was of the order of 1 ruble per man-day in 1962 and two-thirds of a ruble in 1964. As the average money wage of State farm workers in 1964 was 70.6 new rubles per month or roughly three rubles per man-day, daily earnings of collective farmers may have been of the order of $2\frac{1}{4}$ rubles or about 80 per cent of the wages of

State farm workers and just over half of those of the average industrial worker.

However, there were wide regional differences which partly account for the variations between individual commodities. In the Russian Republic, collective farm earnings per man-day were the same as State farm wages on grain farms in western Siberia; elsewhere the difference varied from less than 0.4 rubles in the Far East and the north Caucasus to more than one ruble in north and central Russia. In the Ukraine the difference was about one-quarter ruble in the Donets Dnieper region and double this amount in the South, but in Moldavia, the Baltic republics except Lithuania, the Transcaucasus except Georgia, the whole of central Asia and Kazakhstan there was no significant difference between earnings per man-day in the two main types of public agriculture.

If, on a commodity basis, the differences in earnings tended to be larger for cattle-raising and pig-keeping than for cotton and sheep, this must have been largely due to these regional differences between the north-west and the south-east, as well as to the economics of the respective agricultural activities.

Due to the difference in the number of work days which State farm workers and collective farmers put in during the year, the yearly earnings of collective farmers may have varied from less than half the average State farm wage to over two-thirds according to the region and farm concerned. If these figures are anywhere near the mark, they probably reflect a considerable rise in the relative importance of the collective farm as a source of income for its members compared with the situation even a few years earlier. Thus Jasny estimated that in 1958 the collective farm provided the peasants only with a quarter of their incomes, with three-quarters coming from the private plots.[63] Such estimates cannot be made with any degree of precision and are inevitably subjective, because the scanty figures on the private economy of the agricultural population cannot be divided between collective farm peasants and other rural workers.

The agricultural activities of the 'population' accounted for 45.4 per cent of total agricultural gross production in 1953, 37 per cent in 1958 and 33.1 per cent in 1962. They may have increased by something over 30 per cent between 1953 and 1958 and fell by 4 per cent between 1958 and 1962.[64] They declined further by about 4 per cent between 1962 and 1964, when they may have amounted to about 29 per cent of total agricultural gross production.[65]

Little or nothing is known about the income in kind drawn by the peasants from their plots, but it must have been substantial, because the proportion of material inputs in their subsidiary economy was probably lower than in public agriculture and they supplied a much lower proportion of marketable surplus than did public agriculture. It is, however, possible to divide peasants' cash receipts from market transactions into receipts from public trade and from the collective farm market by deducting the revenue of the collective farms as such from the cash value of Government procurements (excluding State farms) and from the total value of collective farm trade.

TABLE 53

Cash receipts from sales from individual plots 1952-64

	(i) ·000m new roubles			(ii) 1952=100			
	Public trade*	Coll. farm trade	Total	Public trade	Coll. farm trade	Total	Public trade in % of total
1952	0.68	3.90	4.58	100	100	100	15
1958	2.56	3.29	5.85	377	84	127	44
1960	3.28	2.84	6.12	482	73	133	52
1962	3.91	4.04	7.95	575	103	173	50
1964	3.70	3.60	7.30	544	92	159	51

* Public trade=Government procurements *plus* co-operative trade.
Sources: 1952-62: Narkhoz 1962, pp. 240, 342, 540. 1964: Narkhoz 1964, pp. 257, 400, 657.

Though the margin of error in global estimates of this kind must be high, in normal times the near stagnation of the peasants' cash returns from the collective farm market must have greatly reduced the economic importance of this source of income. There are, however, two significant exceptions to this broad conclusion: in 1962 receipts from the collective farm market reached a record, exceeding even the 1952 level, when conditions were much more favourable to the collective farm market than in later years; the 1964 results show a drop in receipts from public trade, both absolutely and in relation to total cash receipts. The 1962 cash receipts from the collective farm market must have been boosted by the increase in shop prices of meat and butter in the summer of that year, while the supply of livestock products in 1964 was affected by the 1963 crop failure. The effect on receipts from public trade procurements in that year may be fortuitous, because these were particularly high in 1963.

5. THE REORGANIZATION OF AGRICULTURAL ADMINISTRATION

In a bureaucratic régime such as that of the Soviet Union policy choices assume in the first place the form of adminstrative reorganizations. The shape and functions of the top management of the agricultural system under Khrushchev were partly determined by the changes in the structure of economic management in general and partly by the changing fortunes of his agricultural policies. Here, too, there is a fairly clear-cut dividing line between the 'reform period' 1953-58 and the 'regroupment period' 1959-64.

Agricultural Reform 1953-58
With some over-simplification it may be said that the guiding principle of agricultural organization at all levels during the first and most successful years of the Khrushchev era was the decentralization of management and the reduction in the extent and intensity of central control.

The administrative superstructure was affected by the general stream-lining of the Soviet Government which began within a few days after Stalin's death, when a number of ministries with purely administrative functions were amalgamated into groupings of wider significance. This involved, in the first place, greater centralization, for the Ministries of State Farms, Cotton Growing, Forests and Procurements were merged with the Ministry of Agriculture and Procurements. The new body was supposed to be responsible for the whole of public agriculture and for the all-important organization of agricultural supplies for feeding the towns and supplying processing industries with their raw materials. (March 15, 1953.)

However, this change came almost immediately under fire

from influential quarters. Khrushchev's famous speech on agricultural policy in September 1953 sharply attacked the central control of the ministerial bureaucracy over the productive units —a theme which was steadily deepened and at least partly translated into practice, first in agriculture and not long afterwards in the economy as a whole.

Khrushchev was not content with the ritual of criticizing the unwieldy bureaucracy in general terms and of denouncing the bureaucracy of the Ministry of Agriculture and Procurements, its grotesque over-centralization and its weak links with the practical work in the fields. He contrasted the total employment of 350,000 agricultural specialists with the mere 18,500 on the collective farms themselves and 50,000 in the Machine and Tractor Stations. On the other hand, 75,000 experts were employed on administration and not nearly close enough to the process of production of the collective farms.[66] In the belief that the weakness of public agriculture in trained and reliable managers and technicians was one of the main causes of its poor performance, Khrushchev was determined to transfer the bulk of the chair-borne agricultural specialists to local productive work. However, this was impossible without a radical reduction in the detailed control exercised by the Ministry of Agriculture over the daily life of the collective farms. (The limits of this diagnosis were shown up by the course of events; almost nine years later—when the Ministry of Agriculture was only a shadow of its former self—Khrushchev still talked about the mobilization of agricultural specialists for productive work as a task for the future.[67])

The re-establishment of an independent Ministry of State Farms in September 1953 and of the Ministry of Procurements in November of the same year were less important in Khrushchev's decentralization policy than the measures taken to strengthen the 'cadres' of the MTS and the State farms, the local productive institutions of public agriculture in the technical field, at the expense of the central bureaucracy. In the autumn of 1953 the Government closed the local offices of the Ministry of Agriculture which until then had been 'the most important planning, control and verification organization in rural administration'.[68] At the same time, the district Party organization

211

became directly responsible for the supervision of the MTS and, through them, for the operation of the collective farms in its area.[69] This change was, of course, connected with the power struggle in the hierarchy between the First Secretary of the Communist Party, whose power base was the Party organization, and the Prime Minister who was primarily in charge of the Government bureaucracy.

The reorganization of the central system of agricultural organization took a little longer and was enacted only in March 1955 by a Decree of the Party Central Committee and the Soviet Government. This was very critical of the tutelage of collective farm operations by the Ministry of Agriculture for interfering with the initiative of the local productive enterprises and ruled that in future the central authorities should confine themselves to fixing obligatory delivery quotas and the semi-commercial Government procurements in excess of the quotas, to determining the services of the MTS to the collective farms and to laying down the rates for payment-in-kind for these services. Within the narrow limits left by central decisions on these vital issues, collective farms, MTS and State farms were to be free to plan their own production.

The reorganization of agricultural administration was the precursor, and may have been the model, of the great upheaval of May 1957 when Khrushchev attempted to destroy the power of the central ministerial bureaucracy over the economy by liquidating a large number of central economic ministries and transferring the power of higher economic managements to more than a hundred regional Economic Councils. This was a far-reaching and not very well considered reform which brought the opposition to Khrushchev in the Party to a head and almost led to his overthrow.

Although not nearly so radical in practice as in theory, this was the first major change in the bureaucratic framework within which Soviet agriculture had operated, by and large, since the mass collectivization of the early 1930s. It was accompanied by substantial changes in the balance between State farms and collective farms and followed soon afterwards by the abandonment of the system of controlling the collective farms through the MTS.

212

The changes made in the MTS in 1953 soon after Stalin's death, did not suggest that the authorities were contemplating any such far-reaching measures. On the contrary, the transfer of large numbers of staff from the collective farms to the MTS in 1953 and the replacement of political deputy directors by secretaries responsible to the local Party organization indicate the opposite. However, after the relaxation of detailed official tutelage over the operations of the collective farms in 1955 and the abolition of detailed output schedules, the political and economic control of the MTS over the collective farms in their area of operations became something of an anachronism. Tension between collective farm chairmen and MTS directors mounted with the growing status and power of the collective farm management. 'Dual control' became intolerable as soon as it tended to become a reality, and the trend of Khrushchev's farm policy —as, indeed, of his economic policy in general—at the time made it inevitable that the conflict would be resolved in favour of decentralizing the system as far as possible, by doing away with the MTS. The official purpose of this radical measure was to give more elbow room to the development of agricultural production by uniting the control of the land and the machinery needed for large-scale cultivation; it was expected that this would permit the collective farms to raise output more effectively because it would make for much greater local initiative.[70]

Another motive for the change was the fact that the traditional system of paying the MTS in kind for their services interfered with the prevailing trend of putting the relations between the régime and the peasants on a simple cash basis; this should encourage the peasants to produce more and to sell more of what they produced at attractive prices to the public trade system. It would, no doubt, have been possible to replace payment in kind to the MTS by cash payments, but if these were to be adequate to cover the cost of these services in full, they might well have been so high as to become a source of further friction. This would have been the reverse of the high real cost of the products obtained through payments-in-kind to the MTS of which Khrushchev had frequently complained.[71]

On a different plane, the winding up of the MTS and the transfer of their machinery to the collective farms at a price

213

which would claw back some of the extra purchasing power given to the peasants through increased producer prices, would make future investment in the collective farms even more than in the past a matter for the peasants themselves. The State tended to concentrate agricultural investment during this period on the 'new lands' where State farms predominated, and the growing reliance of these areas for grain supplies may have encouraged the belief that it was safe for the Government to pull out of collective farm management.

It could be argued that the reasons adduced by Stalin against handing over the MTS to the collective farms in 1952 had lost much of their force by 1958. The growing share of the State farms in market supplies disproved the fear that, without control of the MTS, the régime would become too dependent on the peasants. At the same time, the growth in the income of the collective farms through higher producer prices lent some support to the view that the acquisition of the main items of mechanized equipment by the collective farms need not interfere with the modernization and mechanization of agriculture. Whatever the logical force of Stalin's reasoning, on this point he was more nearly right than his more optimistic successors.

Be this as it may, once the authorities had made up their mind on the matter, the bureaucratic machine lost no time in the execution of the design. The proposal to wind up the MTS became law on March 31, 1958, and it was immediately put into force with all the formidable powers of persuasion and pressure at the disposal of the régime.

After eight years of collective farm amalgamation and a year or two of conversions into State farms, the numerical discrepancy between collective farms and MTS had somewhat narrowed. At the end of 1950 there were 8,414 MTS for 121,353 (agricultural) collective farms, a ratio of almost one to fifteen; at the end of 1957 there were 7,903 MTS for 76,535 collective farms, or more than one to ten.[72] However, there were wide regional variations: in such important areas as the central black earth zone, the Volga region, the Urals, the northern Caucasus, Siberia and central Asia, i.e, in the new and traditional centres of wheat cultivation and in the main cotton-growing districts, there were on average only five to seven collective farms per

214

MTS, while in the grain deficit areas of the north and north-west and in Transcaucasia the proportion may have been as high as fifteen or twenty to one. This was apparently due to the fact that in such regions animal husbandry and horticulture were much more important than the basic functions of cultivating and harvesting field crops.

As a rule there could be no question of a simple and clear-cut transfer of complete MTS assets to single collective farms. The task of dividing staff and machinery was difficult and work schedules—or the structure of field brigades—had to be re-arranged. At first it was the intention to maintain the repair facilities of the MTS in the form of separate Repairs and Tractor Stations (RTS), but this policy was quickly abandoned and in 1961 they were handed over to the district branches of *Selkhoztekhnika* and their staff spread over various organizations.[73] This was less a solution of a difficult problem than a shrugging-off of its existence, and the result was a serious fall in the already low level of maintenance and repairs of agricultural plant.[74]

In the single year 1958 the collective farms acquired about 930,000 tractors (in 15-h.p. units) and 258,000 grain combine harvesters, and this was followed in 1959 by a further transfer of 230,000 tractors and 34,000 combines. The staff employed by the MTSs and their successors declined on average by 1.3m. in 1958 and by an additional 800,000 in 1959.[75]

These figures reflect an organizational revolution of the first magnitude, carried out at breath-taking speed. Nobody can disentangle its consequences on the efficiency of collective farm management from those of the other measures introduced in quick succession, but it is most probable that it contributed on balance substantially to the growing agricultural difficulties of the next few years. As a technical method of ensuring the best possible utilization of scarce equipment, special contract services had obvious advantages over the distribution of heavy tackle amongst individual farms. It is, of course, true that in the operation of the MTS technical considerations had been fused with, and often subordinated to, political aims which had always been problematical and which had become irrelevant during the later 1950s. However, instead of refashioning the relationship

215

between collective farms and MTS on purely technical lines, Khrushchev tried to solve the problem by inverting the political balance between the two institutions and in the process probably weakened the technical efficiency of public agriculture even further.

One of the consequences of the abolition of the MTS was a further impetus to collective farm amalgamation. The MTS had been designed, however imperfectly, for the purpose of providing a large number of relatively small collective farms with modern plant; this avoided the waste of scare resources through under-utilization.[76] If this aim was to be fulfilled after doing away with the MTS, the size of units had to be further increased and their number correspondingly reduced.

With the MTS out of the way, the only remnant of the original 'Stalinist' system of organizing the economic relations between the peasants and the State was the duality of low-priced delivery quotas and normally-priced Government purchases. This vestige of the era of administrative compulsion was out of line with the current policy of relying on the financial self-interest of the peasants and produced quite irrational price relationships. In fact, it was no longer an incentive to increase market supplies and had become an obstacle to smooth relationships in the countryside. It was abolished in June 1958 and replaced by a schedule of uniform prices for all State procurements from collective farms, though with suitable regional differentials.[77]

Regroupment and Experimentation, 1959-64

The administrative reforms of the years 1953 to 1958 were designed to reduce the centralized control of agriculture by the Government bureaucracy in favour of the operational autonomy of the basic units of public agriculture which could be either collective farms or, to a growing extent, State farms. This policy was, in principle, not confined to agriculture but at that time it was applied to other sectors of the national economy only at the top through the abolition of central ministerial control in favour of regional management bodies (*sovnarkhozy*) which, incidentally, created two new problems for every one it solved.

In agriculture, on the other hand, the main aim of official policy was the rationalization of decision-making on the large-scale farms of public agriculture, which were to be left reasonably untrammelled by overlapping and competing influences. At the same time they were to be controlled in key decisions for the régime by the local representatives of the Party machine. The district secretary of the Communist Party was in the last resort responsible for ensuring that official policies—which were ultimately the instructions given by Khrushchev and his immediate advisers—were carried out throughout the length and breadth of the country.

This method of organizing agriculture was incompatible with the rival system of a powerful Minister of Agriculture in executive control of the collective farm economy through his own bureaucracy, and a parallel Minister of State Farms for most of the rest of public agriculture. The All-Union Ministry of State Farms was disbanded as early as 1959 and replaced by regional (Union-Republican) ministries, where the number and importance of the State farms warranted this; the Ministry of Agriculture was deprived by 1961 of all its executive functions in the fields of supply and finance (and of the forceful leadership of its head, V. V. Matskevitch) and removed from Moscow, as Khrushchev regarded it as wrong that the administrators of agriculture sat 'on asphalt' rather than 'on the soil'.[78]

The grass roots reorganization of public agriculture and the permanent friction between Governmental and Party bureaucracy made a reform of the traditional system essential but it did not do away with the need for direction and control. Decentralization had taken place in a period of rising production which appeared as the most telling argument in favour of a policy which was accompanied by such good results. The organizational problems to which it gave rise became fully apparent only during the years of stagnation after 1958, and the disappointments of these years drove the top leadership into increasingly hasty and ruthless agricultural and organizational decisions. Finally, in the course of 1962 to 1964, its reaction took on a tinge of exasperation and produced results which caused widespread dissatisfaction amongst powerful interests.

On the purely technical level, the disbandment of the MTS, the emasculation of the Ministry of Agriculture and the decentralization of decision-making created the need for a new central service organization at both ends: the supply of public agriculture with its growing requirements of industrial inputs such as fertilizers, plant and machinery and the marketing of farm produce in the absence of the administrative compulsion embodied in compulsory production quotas. The need for a new supply organization was met, after some administrative fumbling, by setting up in 1961 the *Soyuzselkhoztekhnika* or Agricultural Technical Association which was to service both branches of public agriculture with plant and spare parts, while somewhat later a separate organization called *Soyuzselkhozkhimia* was created for the purpose of supplying fertilizers and other agricultural chemicals. The Technical Association was a multi-tier body operating centrally and regionally with councils on which leading executives of collective farms, State farms and agricultural machinery plants were represented.

On the marketing side, State Procurement Committees were set up in 1961 whose function it was to reconcile the effects of greater farm autonomy with the needs of the national economy as a whole. These Committees were the most recent incarnation of the frequently abolished, and as frequently resurrected, Ministry of Procurements; like the Supply Committees, they were multi-tier organizations operating on different geographical levels.

Their marketing technique was mainly based on the process of *kontraktatsia*, the conclusion of short-term or medium-term contracts with individual State or collective farms which gave the producers assured outlets for the products most needed by the economy as a whole. Contracts were, therefore, a suitable method of encouraging specialization but in practice they were supplemented by farm inspectors employed by the Procurement Committees. Their primary function was liaison between the Committees and the farms, but they also tended to take an active interest in production plans, although these were nominally a matter for the local farm management alone.

With his usual frankness, which made him the *enfant terrible* of the Soviet leadership but which was probably one of the qualities responsible for his personal pre-eminence, Khrushchev contrasted the legal autonomy of the 'producers' co-operative' in the choice of its chairman and the arrangement of its production plans with the fact that in practice the authorities had to interfere with both, because agricultural prosperity was fundamentally important for the prosperity of the people.[79] From this point of view what was needed was a new and more efficient organ of management of the system of public agriculture in both its forms: 'Institutions which carry out the overall direction of agriculture we have in plenty, but a body which would manage agriculture, look after the organization of production and procurements, grasp thoroughly the need of the State farms and the collective farms, direct in detail the development of every unit, attempt to improve the effective output of the soil—such a management body we have not got, nor has there been one, in essence, throughout the years of Soviet rule. Agriculture has been, and remains, under-managed.'[80]

In an emergency, local bodies for supervising the extent to which national policies are carried out on individual farms may be needed even in countries which normally pride themselves on their free economy; of such a kind were the County War Agricultural Executive Committees in Britain during the Second World War, which had important administrative and supervisory functions to ensure that output from the land was increased as much as possible. Under the Soviet régime such a need is both more obvious and more permanent, particularly in view of the co-existence of State farms and collective farms and of the limited rôle of the market in the allocation of resources.

The solution of this problem adopted by the Soviet Government in the spring of 1962 was the creation of Territorial Production Administrations (TPA). Together with corresponding bodies at the regional (oblast) level, these were the really operative part of a general reorganization of agricultural administration. The top-level tier of the system, an All-Union Committee on Agriculture, remained a paper body. The TPA were to contain representatives of the State and collective farms in their areas and to cover more than one district (rayon), hence their description as 'territorial'. They were intended to take an active hand in the improvement of farm production through 'inspector-organizers' who were to advise farm management on the most progressive techniques and to report to their parent bodies, if State farm directors or collective farm chairmen refused to take their advice. The system was first designed on the basis of 800 to 1,000 TPA and bore some resemblance to the regionalization of economic management through the regional *sovnarkhozy* in 1957.[81]

Despite the fact that representatives of the local Party and Government hierarchy were to sit on the new committees, the whole arrangement was a radical departure from the earlier policy of using the local Party organization, and particularly the district secretary, as the main organs of central direction and control, for the territory of the TPA covered three to four districts. The resulting confusion was at least partly removed by the abolition of the districts in their previous form as units of Party organization in the countryside. The number of TPA was increased to 1,500 and the area of the new basic Party organization was to coincide with that of the TPA. The chief Party officials were integrated into the new system by appointing them as chiefs of the TPA or of their corresponding Party organizations.

The change was thus not merely a reorganization of the control of agriculture—which would not have caused a great stir outside the circles immediately affected—but a drastic interference with the structure of the Party organization and with the standing of its local leaders. This was politically very important, and the resulting upheaval was compounded by its most spectacular consequence: in the autumn of 1962, the Party organization itself was split into separate branches in charge of industrial and agricultural affairs respectively.

In the system of Soviet society, the Communist Party performs the key function of supervising the way in which the Government bureaucracy controls the population. Thus it has a monopoly of genuine political action which it can maintain only because it is a unitary force. When Stalin violated this principle by permitting and fostering the rise of the Secret Police to independent power within, and therefore over, the Party, this all but killed it and endangered the survival of the régime as a whole.

Khrushchev's division of the local Party machine, though by no means so sinister, was almost equally dangerous to the power of the Communist Party; sooner or later it would have prevented it from performing its essential political function which requires that it should remain independent *vis-à-vis* the social classes and the machinery of government. Only a ruler with an independent power base and with a succession of glittering triumphs could have hoped to carry off such a coup. As it was, it only needed the catalyst of failure to precipitate the growing discontent of the Party organization with its energetic and versatile but unpredictable and well-nigh uncontrollable head. When the breakdown of Khrushchev's international policy after the missile crisis of October 1962 and the agricultural calamity of 1963-64 supplied a rich harvest of failure, his fall from power was only a question of time.

It is only fitting that the occasion for the final showdown between Khrushchev and his increasingly exasperated colleagues is said to have been another scheme of agricultural reorganization in October 1964. This involved the creation of seventeen agricultural regions and may have embodied a scheme for setting up separate bodies to manage the production of individual commodities which Khrushchev had proposed two months earlier. The Praesidium of the Communist Party refused to accept Khrushchev's plan and referred it to a meeting of the full Central Committee where Khrushchev found himself, unlike seven years previously, in a minority.

6. THE KHRUSHCHEV ERA IN PERSPECTIVE

If ever the name of an individual can be fairly used as a symbol of an era, it is Khrushchev's in relation to Soviet agriculture between 1953 and 1964. His authority for the many changes made during this period was much more than nominal, and his responsibility for its successes and failures was in a very real sense personal. As a national leader he benefited greatly, and perhaps unduly, from the upswing in agricultural production during the years up to 1958, and his prestige suffered disastrously, and somewhat unfairly, from the crop failure of 1963.

Perhaps his greatest service was to convince the country and his party that Soviet agriculture was a literally vital branch of the economy, and that the policy of starving it of resources and keeping peasants' incomes down was, in the long run, a threat to the prosperity of the Soviet economy as a whole. In his homely way he succeeded better than anybody else in explaining that what happened on the land was of direct concern to every single citizen of the country, because what was at stake was a choice between scarce or plentiful food and clothing for everybody.

However, the task of redistributing both current income and capital investment in favour of the peasants and of public agriculture was much greater than anybody anticipated, and the resistance of the interests adversely affected became increasingly more difficult to overcome. Competing claims on the investment fund raised extremely complex problems of priorities; the enormously costly nuclear armaments programme, compounded by the even more costly space race, jostled with the crying need for more housing in town and country and with the claims of productive investment in industry and transport as well as in agriculture. The substantial rise in agricultural investment during 1953 to 1958 was, therefore, obtained in the teeth of strong opposition from other claimants and had to be limited to the bare minimum promising to produce quick results.

Similarly, the long overdue redistribution of current income in favour of the peasants and agricultural workers through increases in the prices at which the Government bought the produce of State farms, collective farms and, to a lesser extent, of the collective farmers, involved the authorities in growing friction with powerful interests. The earliest, and in a way the most seriously affected, sufferer was the public exchequer.

In the last years of the Stalin era, and during the uneasy coalition period immediately afterwards, the annual reduction in consumer prices in early spring had been a gesture of political importance: it was an earnest of better times to come after a hard winter, tangible evidence for the urban consumer of rising living standards, another step on the long road of reaching and overtaking the material level of Western Europe and the United States. Although this special policy was abandoned after 1955, food prices in public trade as a whole fell by about one-eighth between 1952 and 1958 and remained, on balance, at this lower level till near the end of the Khrushchev era. These price changes, modest as they were, generally involved a genuine transfer of resources in favour of the urban consumer. Price reductions in the collective farm market—which followed the level of prices in the public trade system in normal years—were in effect made at the expense of the peasants, but lower prices in the very much larger food turnover of public trade involved a substantial drop in the share of the third partner—the Government.

In 1958 ordinary bread was almost 20 per cent cheaper than it had been in 1952; its price remained unchanged throughout the rest of the Khrushchev era, although the Government paid the producers more than seven times the absurdly low grain prices in force under Stalin. Vegetable oil had become relatively even cheaper, although sunflower seed returned the peasants almost ten times as much as in 1952; even where retail prices showed some increases, these were only a fraction of the increases in procurement prices.

Although the yield of the turnover tax rose somewhat during the post-Stalin period in absolute terms, it declined sharply in relation to retail turnover. In 1952 sales through public trade amounted to 39,359m. (new) rubles at retail level and turnover tax yielded 24,689m. rubles.[82] By 1958 retail turnover had risen to 67,720m. and turnover tax to 30,454m. (new) rubles. Six years later, retail sales through public trade had expanded further to 96,361m. and turnover tax to 36,694m. new rubles.[83] Between 1958 and 1964 the rate of turnover tax on heavy industry rose slightly, from an average of 6.7 per cent to 6.9 per cent of wholesale prices including tax. However, during the same period the rate of turnover tax on light and food industry output fell from 33.7 per cent to 26.3 per cent. Removing the optical effect of expressing turnover tax as a proportion of wholesale prices including tax, this means that products of the light and food industries were charged with turnover tax at an effective rate of about 50 per cent on factory prices in 1958 and of just over one-third in 1964.[84]

After the unification of producer prices in 1958, processing and distribution margins of some foodstuffs seem to have absorbed most or all of the difference between producer prices and resale prices, particularly for livestock products;

otherwise the Soviet leaders would probably have refrained from raising retail prices of meat and butter in June 1962 despite the loss of face and popularity which this undoubtedly involved. It is reasonable to conclude that the previously almost unlimited room for manoeuvre provided by the indecently wide gap between producer prices and consumer prices had been considerably reduced by the operation of economic forces which the Government did not dare to ignore any longer.

The shock effect of this development was particularly strong, because the improvement in the relative economic position of the agricultural population had already caused unmistakable resentment amongst other classes about the distortion of the generally accepted income relationships between industrial workers and peasants.

The impetus imparted to agricultural output through larger investment and higher producer prices could not be maintained indefinitely by larger doses of the same stimulants. This was the more serious, because the injection of new resources, though respectable by any standard, had not been sufficient to offset the effects of the great economic blood-letting of mass collectivization and war on the capital assets of agriculture, and even the vastly improved producer prices continued to lag behind the level of production costs for important commodities such as livestock products. Khrushchev had managed to lift agricultural production out of the rut of the last years of the Stalin era, but not to create the conditions for self-generating further progress.

Nothing illustrates more clearly the inherent dilemma of the whole epoch than the development of the 'new lands' which was both the most spectacular success of Khrushchev's leadership and its Achilles' heel. As an attempt to buy time during which a long-term programme of agricultural reorganization through more intensive land use could gradually unfold, it was a bold and imaginative concept, carried out with enormous bustle and energy and, perhaps, not more inefficiently than was inevitable in the circumstances. As an alternative to a considered long-term policy of intensification it was bound to fail. Yet it was not until its weaknesses were already painfully evident in the fall of yields and the threat of a huge dust-bowl that the intensification of agriculture in the settled areas was tackled with the necessary resources. The great campaigns in favour of maize and leguminous crops, though not without intrinsic sense in some parts of the country, were in certain respects essentially 'gimmicks', attempts to obtain the fruits of intensification without due investment through the windfall of large increases in yields through changes in cropping patterns.

The great setback of 1963 was largely due to a particularly bad season, but it was also the culmination of a process which had been going on since 1959—a growing disproportion between resources and results obtained. Gross production per hectare on collective farms and State farms increased only from 62.8 rubles in 1959 to 66.0 rubles in 1962, but current expenses per hectare shot up from 44.3 to 55.4 rubles. Thus agricultural gross income per hectare declined from 18.5 rubles in 1959 to only 10.6 rubles in 1962.[85] The shadow of failure hovered over Khrushchev's agricultural policy well before it became manifest in 1963.

Some of the structural changes which distinguished the Khrushchev era were already well under way at the time of Stalin's death, particularly the amalgamation of collective farms with which Khrushchev was, of course, closely associated from the start. The whole period is, however, characterized by tremendous activity in every field. The system of public agriculture, the administrative arrangements for its supervision and control, the enterprise pattern between regions and types of farms, all experienced more or less radical upheavals. Apart from their—temporary or lasting—effects on the operating efficiency of agriculture, they tended to disguise the crucial fact that the really important measures initiated by Khrushchev, and prosecuted with all the energy of which he and the bureaucracy under his control were capable, were essentially short-term palliatives. They were undoubtedly necessary, most of them were by no means ill-conceived and they had a considerable measure of success, but a systematic long-term policy of replacing the inefficient, low-yield system inherited from Stalin by an intensive, high-yield agriculture was missing. Indeed, even its necessity was proclaimed with due emphasis only after the emergency measures had lost their original impetus, and when its execution had become correspondingly more difficult.

Perhaps the greatest weakness of Khrushchev's policy was thus his failure to look beyond the—admittedly pressing—needs of the day and to retain a sufficiently realistic attitude towards the expedients he employed with such energy and ingenuity. In this cardinal respect he reflected the limitations of the bureaucratic system within which he operated. The results of this crucial weakness can be seen in their most extreme form in the tasks assigned to agriculture in the Seven Year Plan for 1959 to 1965, which was based throughout on the faulty assumption that the expansion in output obtained during the preceding five years would be fully maintained and steadily extended during the foreseeable future.

The average yield of grain in 1958 was 11.8 quintals per hectare—barn yield, it was claimed at the time, but in fact bunker yield. This was not a very impressive performance compared with Western experience, but a peak yield nonetheless, and the first year in which an average yield of more than ten quintals per hectare had been obtained. The Seven Year Plan, however, regarded this not as a windfall but as the starting point for a great leap forward which would raise average grain yields by three to four quintals per hectare by 1965.[86] In fact, the 1958 yield was not equalled until 1964, and the average for the seven-year period fell considerably short of the 1958 level.

This crass failure cannot be attributed to the shortcomings of the fertilizer industry. Although it did not—and could not—reach the wildly excessive targets set by Khrushchev in 1962, it fulfilled the original 1965 target of 35m. tons[87] only one year after the due date and exceeded 30m. tons in 1965.

Most of the other agricultural production targets of the Seven Year Plan were no less out of line with reality than those of grain production and seem to have been based frequently on crude extrapolation of the rate of progress between 1953 and 1958. The outstanding exceptions were raw cotton and sugar-beet, both of which performed broadly in line with the demands of the Plan. On the other hand, potatoes—Russia's 'second bread' and rightly regarded as a key crop because of their double use as human food and animal feed—failed completely to expand and were barely higher in 1965 (or in 1964-66) than they had been in 1958. Consequently, the expectation of deriving an extra three million tons of pig meat from this source alone[88] proved unfounded.

No less abortive was the whole animal husbandry programme which was completely disorganized by the effects of a single serious crop failure. Instead of producing 'not less than' 16m. tons of meat in 1965 (compared with 7.7m. tons in 1958), actual output just exceeded 10m. tons. Milk production rose from 58.7m. in 1958 to 72.6m. tons in 1965—a far from negligible rise but well below the 100-105m. tons postulated by the Plan[89] and egg production, after rising from 23,000m. in 1958 to just over 30,000m. in 1962, did not exceed this level again until 1966, when it still remained far behind the Plan figure of 37,000m. for 1965.[90]

This confrontation between Plan and reality would hardly be worth while if it were simply intended to prove the all-too-well-known fact that agricultural production plans in the Soviet Union were much more the expression of aspirations than either projections of genuine trends or reasoned expectations that specific measures would produce specific results. For on this occasion, more than at any other time in the history of Soviet agriculture, specific and positive measures had, indeed, been taken. Some of them, but by no means all, may have been ill-considered, all of them suffered severely in their application from the continuing backwardness of the country's agriculture and from the bureaucratic defects of the whole system; more important still, despite the enormous effort which had gone into them, they were simply insufficient for their purpose. To have willed in great things may have been enough for the resigned members of a decaying world, but it did not satisfy the mass of Soviet consumers who, after long years of deprivations, wanted more solid fare.

IX

After Khrushchev

Khrushchev's rise to power may not have owed much to his reputation as the foremost agricultural expert in the Communist hierarchy, but his downfall was almost certainly hastened by the hectic, confusing and at times contradictory measures by which he tried to carry out his agricultural policy after 1958. The surly criticism of his subjectivism, his ill-considered ('hare-brained') schemes and his disregard of technical advice, of which his exasperated colleagues accused him correctly enough, was remarkably mild by the standards of Soviet polemics; it also differentiated clearly between the personality of the discredited leader and the policies which commanded a broad measure of support by the leading group. The new rulers did not condemn his basic policies to which they themselves were committed but his style of leadership, his disregard of technical limitations and his bureaucratic emphasis on administrative action, though the latter was deeply ingrained in the whole system of Soviet government and not attributable to any individual.

Brezhnev and Kosygin, who emerged as the political and administrative top representatives of the Soviet establishment, consolidated their independent position through the retirement of Mikoyan from the presidency where he may have played a key rôle in facilitating the transfer of power from Khrushchev to the new leadership. If they were content to have rid themselves of an exuberant and unpredictable chief without trying to label him as an 'anti-Party' element, this may have been simply due to the fact that there really were no serious policy differences between the old boss and his one-time assistants. Thus they contented themselves, on the whole, with carrying out certain measures which were widely accepted as necessary but which had been in danger of becoming discredited by the bureaucratic extremism of their execution.

The crop failure of 1963 and its effects on livestock production had created an emergency in which the main emphasis had to be placed on steps which promised a quick recovery in production. In the short run, this meant above all more incentives for the peasants, both in their capacity as collective farm members and as cultivators of their subsidiary private plots. In the longer term, the Government had to come to grips with the ingrained weaknesses of the agricultural economy, to re-define the methods by which output could at last be raised to the level demanded by the needs of the country, to allocate the necessary resources and to find a new solution for the chronic administrative defects of public agriculture, the lack of a coherent policy in its central administration and the continuing weaknesses of its main forms, the collective and State farms.

1. BACK TO 1958

The new leaders were eager to brush aside the problems created by the unsuccessful expedients adopted by Khrushchev during his later years and to consolidate the advances made in the first five years after Stalin's death. The immediate need was for a new impetus capable of overcoming the impasse of 1963-64 and to gain time for modifying some policies and abandoning others whose success had not matched the expectations originally put into them.

The most ruthless and controversial changes made by Khrushchev after 1958 were in the management and control of agriculture by Government and Party. Here the new masters acted immediately and with a clear bias in favour of the traditional system. One of their first decisions was to re-establish the unity of the local Party organization which Khrushchev had divided into an agricultural and an industrial branch despite unmistakable friction and dissatisfaction by the Party machine. Similarly, the emasculation of the Union Ministry of Agriculture and its removal from Moscow were quickly reversed, and in February 1965 Khrushchev's *bête noire,* its energetic head V. V. Matskevitch, was reinstated after a period in exile in the new lands. It indicates the individual style of the new régime that Matskevitch's successor Volovchenko was not made to suffer for the eclipse of his patron but was simply kept on as second-in-command. A similar change was made in Kazakhstan, and Kunayev was re-promoted from the position of Prime Minister to his earlier post as First Secretary of the Communist Party. Later in the same year the 'virgin lands territory' constituted by Khrushchev was again divided into its constituent administrative regions. The Territorial Production Administrations, the most important operative part of the administrative reorganization instituted by Khrushchev, were realigned with the local administration boundaries in a reduced number of districts (*raions*).

But administrative reform, however important in itself and particularly within a bureaucratic system, was not the key to an improvement in the immediate agricultural crisis. This needed an increase in resources devoted to agriculture. Such an increase had been the main cause of the considerable successes achieved up to 1958 and the interruption of this process during the following years had been too closely associated with the relapse into stagnation to be explained away as fortuitous. Bigger investment allocations and bigger incentives were regarded as necessary, and further increases in producer prices were, therefore, made almost immediately.

The earliest product to be reviewed was milk. Producer prices were raised from January 1, 1965 and the price structure was radically simplified by abolishing the differential between procurement from State farms and collective farms, reducing the number of pricing regions and removing the premium on supplies for the liquid market.[1] At the meeting of the Central Committee in March 1965, the new General Secretary himself announced revised pricing arrangements for meat and cereals at the start of the new production season, when they could be expected to have the strongest immediate impact. The basic procurement prices were increased by substantial amounts according to region and the unified acquisition prices for 'deliveries' and 'supplementary sales' which had been introduced in 1958 were abandoned; delivery quotas for cereals were reduced to 55.7m. tons in 1965 and were to remain at this level until 1970, and the quota for meat procurements was reduced from 9m. tons to 8.5m. tons, with planned increases thereafter. The earlier practice of varying procurement prices from year to year, and even during the year, in order to stabilize returns despite fluctuations in the size of the crop, had not worked out as expected and had given rise to complaints because it made farm planning more difficult. Longer-term targets were pronounced to be necessary in order to encourage specialization, particularly for livestock products, because herd expansion and the necessary expansion in the feed base had to be planned for several years ahead.[2]

The limitation of delivery targets was financially of great importance for the peasants and the State farms because at the same time substantial price premia were reintroduced on sales in excess of the quota quantity. The limitation of planned deliveries did not mean that the Government wanted to reduce its procurements; on the contrary, the purpose was to stimulate total procurements through the higher prices paid for extra deliveries; this constituted a return to the principles in force before 1958 and, indeed, ever since the creation of the collective farms. However, at the same time the basic purchase prices were also raised quite substantially; for wheat, rye, buck wheat, millet and rice the increases were from 12 per cent to more than 100 per cent, for beef by 20 to 55 per cent, for pork by 30 to 75 per cent, and for sheep and goats generally by 10 to 70 per cent, but even more in mountainous regions.[3] On the last occasion when producer prices for livestock products had been raised, in June 1962, retail prices had also gone up. Now the authorities made it clear that these prices would be left untouched and the cost of the increases had to be borne by the processing plants and the Government.*

The extent of the effect of increases in agricultural producer prices on the profits of the processing industries in 1965, though considerable, was far from disastrous (see Table 54). The major branch of the food industry with a direct agricultural base whose profits were severely reduced in 1965 was the milk processing industry, where the rate of profit had previously been remarkably high, with the result that this industry had contributed a

---

*The extra cost of 2,100m. rubles to the 1965 Budget should not be interpreted as a price subsidy to agriculture but as a consequence of the stepping up of capital investment in agriculture by the Government.

TABLE 54

Profits of the light and food industries, 1962–65
(in '000 m. rubles)

	1962	1963	1964	1965
Light industry	3.07	3.19	3.42	3.47
Food industry	4.44	4.70	5.39	4.64
incl. sugar	0.45	0.33	0.50	0.53
meat	0.43	0.46	0.47	0.70
fish	0.21	0.36	0.54	0.37
milk and milk products	1.13	1.20	1.39	0.57
Total industry	18.59	19.60	21.91	22.55

Sources: Narkhoz 1964, p. 749. Narkhoz 1965, p. 757.

quarter of the profits made by all branches of the food industry.

Increased producer prices were not the only source of higher net returns to agriculture after Khrushchev's fall. Prices of industrial goods supplied to agriculture were reduced and taxes cut. The tax on collective farm incomes was reorganized by the decree of April 10, 1965, which excepted not only the 'normal' wage payments to collective farm workers and contributions to the Social Security Fund but also an average profit of 15 per cent from this tax which thus largely assumed the character of a rent payment[4] and whose amount was roughly half of what it had been before. The State farms were also permitted to retain a substantial proportion of their profits.

Another well-tried emergency measure for a quick increase in market supplies was used on the same occasion; this was the loosening of restrictions on the subsidiary private economy of the rural population. Concessions on the size of livestock holdings were made almost overnight and early in 1965 the tax on private livestock was dropped, price restrictions in the collective farm markets were lifted, the sale of produce at railway stations and river crossings was again permitted and plans to build better markets were announced.

It remains to be seen whether this policy is more than a short-term expedient, as it was immediately after Stalin's death, for the authorities have an ingrained reluctance to accept the extension of a market over which they have only limited control. In any case, in 1965 the subsidiary peasant economy raised its output by almost 8 per cent and was only a shade lower than it had been in 1958,[5] while private livestock holdings which had been on the decline since 1958 started to rise again despite the effects of the harvest failure of 1963. However, the fact that this process was again reversed in 1967 may indicate that official attitudes towards the subsidiary private plots are again reverting to normal.

TABLE 55

Private livestock holdings, 1959–68 (m.)

January 1	Cattle incl.	Cows	Pigs	Sheep and goats
1959	29.2	18.5	15.1	36.4
1965	25.1	16.2	14.5	30.5
1966	27.9	16.6	18.2	32.2
1967	29.3	17.1	16.5	33.3
1968	28.4	17.1	13.6	33.5
1969	27.3	16.7	12.8	34.3

Sources: Narkhoz 1964, pp. 353f. Narkhoz 1965, pp. 368f and provisional reports of the Central Statistical Board on 1966 to 1968.

At least for the time being the Government recognized the fact that the individual contributions to agricultural production remained essential for the supply of potatoes and vegetables, meat, milk and especially eggs; although the major part of the output of the subsidiary plots was destined for consumption in kind, individual plots provided 9 per cent by value of the market produce of crops and 19 per cent of that of animal products in 1964.[6] The bureaucratic dislike and disregard of this contribution to the supply of the urban population had to be curbed and it again became respectable to remind the public that the subsidiary economy of the peasants would lose its economic justification only when public agriculture had reached the point of satisfying the whole national requirements of agricultural produce.[7]

2. FORWARD TO 1970

The decisions of the Central Committee in March 1965 were of similar importance for the direction of agricultural policy during the next few years as the decisions of September 1953 had been in their time; in particular, they were the foundation of the agricultural part of the Five Year Plan for 1966-70 which was published in draft form in February 1966 as 'directives' of the

233

23rd Congress of the Communist Party.[8] In general terms, the Plan called for the doubling of Government investment in agriculture, with corresponding increases in the supply of agricultural machinery to collective farms and State farms, for better material incentives to the peasants and for an improvement in the organization of agriculture.

Brezhnev had announced as early as March 1965 that investment in public agriculture was to be stepped up from 33,000m. rubles in 1960-64 to no less than 71,000m. This sum is extraordinarily large, not only in comparison with past investments in agriculture but also in relation to the level of industrial investment, for the gross investment in the whole of industry during 1960-64 had been 71,136m. rubles.[9] Current agricultural investment is thus planned to equal industrial investment during the immediate past. Put in another way, agricultural investment from public funds during the current five years is scheduled to equal that during the preceding two decades since the end of the Second World War. Whether these plans will be fully carried out or not, such a programme is by itself a far-reaching departure from the order of economic priorities which have been in force since the late 1920s.

Amongst the specific supply targets of the current Plan is the stepping-up of new tractors (to 1,790,000), lorries (to 1,100,000), trailers (to 900,000) and grain combine harvesters (to 550,000); this would involve, by and large, a doubling of the plant inventory at the beginning of the period. One of the main features of the Plan is the extension of the arable area by less risky but much more expensive methods than Khrushchev's development of the new lands: land improvement through the draining of marshes in large areas of the north-west and Far East, the irrigation of dry lands in the south and south-east and the liming of huges tracts of acid lands. A few months after the Congress, a special meeting of the Central Committee decided in May 1965 to double most of these targets—an almost exact parallel to the snow-balling 'new lands' programme eleven years before which is bound to raise doubts about the quality of the staff work behind such flexible targets.

The Plan also recognizes the urgent need for greater usage of mineral fertilizers; its aim of supplying 55m. tons to agriculture in 1970 (compared with 27.5m. tons in 1965) falls far short of the goals set in Khrushchev's crash programme, but may be more attainable for this very reason.

Perhaps in indirect acknowledgment of the fact that Soviet agriculture needs, above all, almost unlimited resources, the Plan is strikingly modest in the setting of production targets to be reached with this massive capital investment, this use of larger current industrial supplies and with substantial price incentives to the producers. One of its features is the expresssion of targets as average results for the period as a whole rather than as specific aims for its final year. Given the still substantial effects of the weather on crops and, with a time lag of up to a year, on animal products, this is intrinsically sound but it makes it more difficult to assess the anticipated rate of growth without simplifying assumptions.

Grain remains the key commodity in Soviet agricultural expansion and the current Five Year Plan calls for an increase in 30 per cent over the average of 1961-65 to 167m. tons. As the 1961-65 average was 130.3m. the target is actually somewhat less than 30 per cent. Though more modest than earlier plans it is still substantial. Ignoring the wide year-by-year fluctuations and taking the good 1964 harvest as a base, a steady increase of 8.5m. tons from year to year, rising to 184m. tons in 1970, would be needed to reach the goal after adjustment for seasonal variations. If this is put against the target for 1970 in Khrushchev's outline plan for twenty years up to 1980 it certainly represents a radical scaling down, as the figure mentioned there for 1970 was 14,000m. pood or 230m. tons. On the other hand, the figure is not dissimilar from that aimed at in the 1959-65 Seven Year Plan for 1965.

The characteristic curve of grain production in the Soviet Union is not a straight line pointing upward from year to year but a step-like sequence of a sharp upswing in a climatically favourable year, followed first by a setback and then by consolidation near the previous peak level as the basis for a further advance. Stagnation during the early 1950s was followed by a rise of almost 20m. tons in 1955 and a further rise of 20m. tons in 1956, when the virgin lands campaign paid off handsomely. A painful setback in 1957 preceded a new record in 1958 which was not again exceeded until 1962. Though all the gains of the previous seven years seemed to be put in jeopardy by the crop failure of 1963, a new record level was reached in 1964 and, after a partial harvest failure in the east in 1965, new ground was broken in 1966, only to be lost in 1967 which was again affected by poor weather in the new lands areas but nevertheless produced a grain crop similar to the 1964 out-turn.

Detailed production targets for other main crops and for animal products were mentioned by Kosygin in his speech at the 23rd Party Congress.[10] Agricultural gross production as a whole should rise by 25 per cent during 1966-70. This is equivalent to an annual growth rate of about $4\frac{1}{2}$ per cent compound—much less than had been demanded on earlier occasions from agriculture, when plans were simple paper exercises in wish-fulfilment, but a respectable task if it is to be realized despite the inevitable setbacks in bad seasons. Furthermore, special weight was put on the need to obtain higher output primarily from better yields, though livestock herds would also go up.

The routine phrases of the official documents about the need of combining central planning with the autonomy of collective and State farms reflect the genuine, if so far inconclusive, efforts of the new leadership to find a better and more stable balance between the interests of the village and those of Soviet society as a whole. This policy is partly expressed in specific measures and partly in the wider search for new methods. The material status of the peasants has been improved by raising their incomes and reducing their expenses and taxes. In addition the Government cancelled collective farms debts to the banks to the tune of one thousand million rubles, while rural retail prices were brought into line with urban prices.

Particularly important was the final solution of the difficult problem of giving the collective farm 'members' a regular income for their work instead of the variable and too-long-delayed labour day 'dividend'. A pilot scheme was set up to explore the possibility of a collective farm wage fund, fed partly from the cash income of the farms and partly from advances by State purchasing agencies on account of future deliveries under contracts. Finally, a guaranteed wage for collective farmers was introduced from July 1, 1966, at rates comparable to those of

State farm workers and based on their 'norms'. This wage is to be paid twice monthly, with special bonuses at the end of the year as a survival of the original concept of a co-operative labour day dividend.

This important reform seems to have given rise to unexpected results: if work for the collective farm is to be paid for at realistic rates it becomes a desirable occupation to be sought out rather than a drudgery to be avoided. Collective farmers therefore may want to do work for their farm even during the slack season, rather than spending as much as possible of their time, however ineffectively, on their subsidiary plots. The current Five Year Plan advocates the setting up of subsidiary enterprises on public farms and inter-collective farm unions in order to occupy the spare time of the agricultural population on similar lines to the rural handicrafts of the past; it is not impossible that the interest of the collective farm members in truly remunerative work may overtake the realization of these sensible plans.

The crucial difference between collective farms and State farms was not the spurious 'co-operative' character of the former but their lack of guaranteed wages to their 'members'. Though the formal conversion of collective farms into State farms has virtually come to an end, the guaranteed wage has factually removed the most important difference between these two forms of public agriculture. The Soviet leaders are obviously searching for a new basis for the collective farms which has become a necessity since the decisions of the Central Committee in March 1965. This is the background to the frequently announced plans for a new Congress of Collective Farmers which would be the first to be held since 1935. A preparatory commission, with Brezhnev himself as chairman, was set up early in 1966 for the purpose of working out a new Collective Farm Statute, based on popular suggestions. Despite the justified scepticism about manipulated popular enthusiasm, the length of time which has been spent on this project is some indication of the seriousness with which the problem has been tackled. The Charter took much longer to prepare than had been anticipated, and the Ministry of Agriculture began to study its draft only in 1967; such delay is an infallible sign of serious differences within the Communist hierarchy which take a long time to resolve.

The State farms which were made financially less dependent than before at the beginning of the Khrushchev era, when they began to be paid reasonably remunerative prices for their products, were also affected by the growing move towards the decentralization of decision-making in other branches of public enterprise. After earlier suggestions that they should be permitted to dispose of part of their profits for the benefit of their staff, a pilot scheme was launched in April 1967 on 390 State farms. Part of the new plan is a 'package deal' combining the abolition of the customary 10 per cent differential between the producer prices received by the State farms compared with those of the collective farms with an interest charge of 1 per cent on the public capital employed by them. In addition, part of their profits is to be allocated to purposes directly benefiting the State farm workers such as bonuses, better housing and social welfare. According to normal Soviet practice, the introduction of such a pilot scheme may be regarded as a forerunner for its general application in the near future; this would, of course, further narrow the gap between the position of the collective farm members and the State farm workers and could be another step towards the complete assimilation of the two types of public enterprise.

3. AGRICULTURAL OUTPUT TRENDS, 1964-68

It has been shown that the changes introduced by Khrushchev's heirs were an attempt to rescue the essentials of an agreed policy from the accretions impairing its effectiveness. This underlying continuity makes it possible to regard the results of the post-Khrushchev era in the field of agricultural production as an extension of the trends established during the previous decade.

Both in order to eliminate the extremes of good and bad years due to weather conditions and in order to reconstruct as closely as possible the terms in which the trend of agricultural production is seen in the Soviet Union, it is advisable to compare current results with the framework of the Five Year Plan 1966-70 and with the average of the preceding five years. (The individual years are shown in the Appendix, Tables II and III.)

TABLE 56

Gross production of main agricultural commodities, 1961–67
(m. tons except eggs and wool)

	Average 1961–65	Average 1966–68	Planned Average 1966–70	Average 1966–68 in % of: (i) 1961–65 (actual)	Average 1966–68 in % of: (ii) 1966–70 (planned)
Grain	130.3	162.8	167	125	97
Raw cotton	5.0	6.0	5.6–6.0	120	100-107
Sugar-beet	59.2	84.9	80	143	106
Sunflower seed	5.1	6.5	—	127	—
Potatoes	81.6	95.0	100	116	95
Vegetables	16.9	19.0	—	112	—
Meat (deadweight)	9.5	11.3	11	122	103
Milk	64.7	79.3	78	123	102
Eggs ('000 m.)	28.7	33.7	34	117	99
Wool ('000 t.)	362.0	39.3	—	109	—

Sources: For all actual results: *Pravda,* January 26, 1969. For 1966-70 Plan: Kosygin's report to 23rd Congress.

In the interpretation of this table full allowance should be made for the fact that the period 1961-65 was one of stagnation in most fields and includes one abnormally bad year for crops (1963) and livestock products (1964). The fine art of statistical sycophancy consists, of course, in the choice of a poor base period for the purpose of putting current results in an unduly favourable light without actually cheating, and the current preference for such comparisons in the Soviet Union certainly owes something to this fact. Nevertheless, the extent and the reasonably uniform rate of progress during the immediate past has been such as to give reason for considerable satisfaction to the Soviet leadership.

Particularly instructive is the comparison between the outturn in 1966-68 and the planned average production for 1966-70. 1966 was a bumper year for grain, 1967 apparently mediocre but not bad and 1968 good for spring-sown varieties; the results of the first three years of the current Plan period make it virtually certain that the Plan will be fulfilled for grain and probably for potatoes and that it will be over-fulfilled to a very substantial extent for cotton, sugar-beet and most or all live-stock products. Whether the Plan was based on an independent assessment of demand in 1970 or on an overcautious projection of supply trends and potentialities cannot be determined from the scrappy information so far available. The cautious and almost timid attitude of the current Soviet leadership indicates the latter; whatever the causes of the under-estimation of output trends, it seems almost certain that, for the first time in Soviet history, the agricultural production plan will be considerably exceeded. It is characteristic that well before the end of the period, a campaign is being organized, e.g. in the dairy industry, to complete the current Five Pear Plan in a shorter period.[11]

One interesting consequence of this changed style in the Soviet leadership from ebullient over-optimism to dyspeptic over-caution is the possible embarrassment which the over-fulfilment of official production plans may cause at the processing and marketing stage. The provisional results for 1967 and 1968 claim that the procurement plans for grain, raw cotton, sugar-beet, sunflower seed, potatoes (but not for vegetables) and for animal products have been over-fulfilled. Though even for vegetable products the need for adequate storage and, in some cases such as sugar-beet, processing capacity has to be met, if large losses at the processing and distribution stage are to be avoided, the position is much more serious for the more perishable livestock products. It may be assumed that the plans for the provision of additional processing capacity were geared to the planned availability of raw material, and it would not be surprising if their execution took more time in practice than the usual optimism of planners and producers allowed. If the actual quantity of agricultural commodities produced at lucrative prices increases much more quickly than planned, the result could be chaos at the processing and marketing end of the Soviet food supply system.

The unrealistically low original targets of the current Five Year Plan were indirectly criticized by Brezhnev in his Report to the Central Committee of the Communist Party in October 1968, when he proclaimed considerably higher short-term aims for the immediate future. These are generally 15 to 20 per cent higher—190-200m. tons of grain, 7m. tons of raw cotton and sunflower seed, 90m. tons of sugar-beet, 115m. tons of potatoes, 14-15m. tons of meat, 90-95m. tons of milk, 45-50 millions of eggs and 480-500,000 tons of wool.[12] Set against the rising trend of the last few years, such targets do not seem outrageously high for a time horizon of three or four years.

Perhaps most encouraging is the evidence for long overdue progress in the improvement of the generally abysmally low yields of Soviet agriculture.

TABLE 57

Unit yields of the main products, 1961–68

Product	Ratio	Average 1961–65*	1966	1967	1968
Grain	qu. per ha	10.2	13.7	12.1	13.9
Cotton	qu. per ha	20.6	24.3	24.5	24.3
Sugar-beet	qu. per ha	165	195	230	263
Sunflower seed	qu. per ha	11.2	12.2	13.8	13.6
Potatoes	qu. per ha	94	105	115	122
Vegetables	qu. per ha	116	125	141	129
Milk	kg. per cow	1,987	2,021	2,128	2,232
Wool	kg. per sheep	2.87	2.89	2.95	3.02
Eggs	eggs per laying hen	132	138	143	152

*Animal products for calendar year 1965, state and collective farms only.
Source: *Pravda*, January 26, 1969.

Perhaps the most significant feature of the development of the livestock industry during the last two years is the check to the cattle and pig herds after their re-expansion in 1965 and 1966. (See Appendix, Table IV.) The cattle herd remained stationary in 1967, with an increase in cow numbers almost compensated by a drop in other cattle, and fell by one and a half million in 1968, when even the cow herd declined by 400,000— exclusively in the private sector. The sharp fall in the number of pigs during 1967 could be ascribed at least to some extent to the effects of foot and mouth disease, but although this factor cannot be ruled out for 1968 as well, it is not impossible that it is at least partly due to a policy of culling unproductive stock and to greater concentration on quality instead of the earlier exclusive pre-occupation with quantity. Whether this interpretation is correct or not, there is little doubt that such action would be amply justified.

241

Though there is obviously a danger of reading too much into the results for a single year, it may be significant that the re-expansion of herds and flocks after the sharp reduction in 1963-64 seems to have slowed down quite appreciably in 1967 (see Appendix, Table IV) and was actually reversed in the case of pigs; as gross production and Government purchases of animal products expanded quite substantially, this may indicate that some of the gain was obtained through culling unproductive stock. Whether this interpretation is correct or not, there is little doubt that there is ample room for such action.

The combination of massive increases in producer returns and in capital investment, good or at least tolerable weather conditions and excessive caution in the setting of production targets thus resulted in output levels in excess of those of the current Five Year Plan. This is, indeed, a unique event in the history of Soviet public agriculture and an indication of the extent of the change in the attitude of the Soviet power towards the peasants, begun but not fully maintained under Khrushchev and resumed by his successors.

PART THREE

Problems and Prospects

X

Open Questions

1. HOW MUCH AGRICULTURAL PRODUCTION DOES THE SOVIET UNION NEED?

After half a century of tight food supplies, punctuated by several spells of severe food shortages and even famine, it is natural to assume that the demand of the Soviet population for agricultural products is to all intents and purposes insatiable. An agricultural policy based on this perspective will be very different from what would be required, if the future demand for food and agricultural raw materials became less voracious and more selective than in the past.

In terms of the quantity of food consumed, the Soviet people with about 3,000 calories available per head of the population in the far from abundant years 1959-61, is amongst the aristocrats of the world's population. There is little room for improvement in this respect, even though the climate of the country and the still heavy reliance of the Soviet economy on manual work may warrant a somewhat higher energy intake than in most other developed countries; protein supplies were also adequate in total, though the share of vegetable protein in the diet was higher than most nutritionists would wish.

The demand projections of the Food and Agriculture Organization of the United Nations for 1975 assume only an insignificant increase in calories per head per day—from 2,998 in 1959-61 to 3,072 on the assumption of a low rate of income growth and 3,087 on that of high economic growth. Similarly, total protein intake is projected to rise only from 85.9 grammes in 1959-61 to 87.5-87.8 gr. in 1975; the main desirable change is a drop in vegetable protein from 56.9 gr. in the base period to 50.2-48.7 gr. in 1975 and a somewhat steeper growth in animal proteins from 29.0 to 37.3-39.1 gr.[1]

245

The total demand for food in terms of energy will thus rise almost exclusively as a result of the population increase, but within this gently expanding total the demand for starchy vegetable products will fall and that for high-protein and protective foods, such as animal products and fresh fruit and vegetables, will go up steeply. These broad expectations have been translated into quantitative projections of global demand for the main types of foodstuffs in 1975 compared with a base period of 1961-63. In the following table these estimates are contrasted with actual production in 1961-63 and with production projections for 1975.

TABLE 58

Selected agricultural products: demand and production (1961–63 and 1975)

	Global demand (m. tons)				Domestic production			
	1961–63	1975 High	1975 Low	Average % rise p.a.	1961–63	1975 High	1975 Low	Average % rise p.a.
Cereals*	95.4	126.3	124.6	2.1	98.5	137.2	127.7	2.3
Sugar	8.0	12.3	11.7	3.2	6.3	12.0	12.0	5.1
Cotton	1.33	1.61	1.53	2.8	1.59	2.37	2.28	6.5
Fats and oils	3.7	6.3	5.9	4.1	3.75	6.2	5.8	3.7
Meat†	7.46	11.6	11.0	3.2	7.46	10.9	10.4	2.8
Milk‡	62.0	85.2	85.2	2.5	63.3	87.1	87.1	2.5
Eggs	1.63	2.4	2.3	2.8	1.62	2.4	2.3	2.8
Wool	0.25	0.38	0.38	3.3	0.23	0.42	0.39	4.6

* Including rice.　　　　† Excluding offals.
‡ Excluding butter which is included in fats and oils.
Source: FAO, *Agricultural Commodity Projections 1975-85* (1967), Vol. I, *passim.*

Within the assumptions on which these calculations are based, it may be said that the demand for cereals may be expected to rise by about 2 per cent compound from year to year, for milk by about $2\frac{1}{2}$ per cent, for cotton and eggs by about $2\frac{3}{4}$ per cent, sugar, meat and wool by about $3\frac{1}{4}$ per cent and for fats and oils by over 4 per cent. Even a modest rise in annual compound rates over such a lengthy period as thirteen years represents a fairly dynamic overall development; at the same time such figures put the size of the demand problem for agricultural products in the Soviet Union into much-needed perspective.

On the production side, the table suggests that in the FAO view production is likely to expand somewhat faster than demand for most crops except, perhaps, fats and oils and in line with demand for milk and milk products and eggs, but somewhat more slowly for meat. Such projections, however carefully made, are inevitably subject to far-reaching reservations which reduce their practical value as forecasts. More relevant than such hypothetical data is the actual development of Soviet agricultural production during the first four years of this period. For this purpose it is unfortunately necessary to limit the comparison to gross production figures which differ in important respects from the definitions used in the FAO commodity projections. However, while this interferes with the direct comparability of absolute totals it does not invalidate the all-important trend comparisons.

The wide fluctuations in most crops make it difficult to devise an entirely satisfactory statistical measure, but the use of a three-year average such as that employed by FAO for the base period is not unreasonable. If the 1961-63 average is again taken as base period and the 1965-67 average as current period, the comparison for some crops—particularly grains—will be somewhat flattering, because of the 1963 crop failure, while that for animal products will perhaps, underestimate the underlying rate of progress to some extent.

TABLE 59

Gross production of selected agricultural production 1961-3 and 1965-7

	(i) in m. tons (eggs '000m units)		(ii) in per cent 1965–67 over		Average rise
	1961–63	1965–67	Increase	1961–63	per year
Grain	126.2	146.7	20.5	116.2	3.8
Raw cotton	4.7	5.9	1.2	125.7	5.9
Sugar-beet	47.5	77.8	30.3	163.8	13.1
Sunflower seed	4.6	6.1	1.5	133.7	7.1
Potatoes	75.3	90.7	15.4	120.4	4.7
Vegetables	15.8	18.7	2.9	118.2	4.2
Meat (deadweight)	9.5	10.7	1.2	113.0	3.1
Milk	62.6	76.2	13.6	121.7	5.1
Eggs	29.3	31.5	2.2	107.5	1.8
Wool	0.370	0.374	—	101.1	0.3

Source: Appendix, Tables II and III.

During the years under review, which were on the whole not a very happy period for Soviet agriculture, production exceeded the long-term growth rate of demand spectacularly for sunflower seed, the most important source of vegetable oil, sugarbeet, cotton and milk. Even for meat, performance was broadly in line with the long-term demand, but for eggs and particularly for wool it lagged seriously behind. Potatoes and vegetables, particularly the former, seem to have expanded satisfactorily, though for vegetables there is a large backlog of demand. Finally, in the all-important case of grains progress was on the surface much greater than required by the underlying demand trend, but this result must be discounted in practice, because the base period included a year of crop failure and the current period one bumper crop as well as a bad year and a reasonably good year.

The data for the current Five Year Plan do not permit an extension of this comparison until 1970, even as far as the plans of the authorities are concerned. This is partly due to the new practice of expressing targets in terms of five-year averages and partly to the over-caution of the targets which were in many cases virtually reached in the first three years of the plan period.

A brave attempt by two American students of the subject to provide their own answer to the question of what the volume of agricultural output in the Soviet Union might be at the beginning of the next decade is worth quoting, at least in abbreviated form (see Table 60).

Although some of the ranges are inevitably very wide, these calculations give a surprisingly optimistic picture. Even if grain production keeps to the lower limit of the range, which would presumably have the effect of limiting meat and milk production to the lower estimates, production would generally rise faster than the long-term trend of demand. This could revolutionize the balance of supply and demand for agricultural products as a whole and could produce substantial export surpluses, though demand may still outrun supplies for fresh fruit and vegetables.

The likeliest candidates for growing export surpluses are cotton, oil-seeds and, at a later stage, sugar amongst crops and certain milk products amongst livestock products. Milk production has been running ahead of demand for some time, though

TABLE 60

Gross production of selected farm products—1961–64 and 1969–72
(in m. tons, except eggs)

	1961–64	1969–72	Average % rise p.a.
Total grain*	132.1	148–175	1.5–3.9
Sugar-beet	55.7	86	5.6
Raw cotton	4.8	6.5–7.5	4.1–5.7
Oil-seeds	5.6	8	4.5
Potatoes	79.7	120–130	5.2–5.7
Vegetables	16.5	25–30	5.3–7.8
Meat†	9.1	12–14	3.5–5.5
Milk	62.7	80–100	3.1–6.0
Eggs‡	28.6	37	3.3
Wool	0.37	0.50	3.8
Flax fibre	0.39	0.55–0.58	4.4–5.1

*According to Soviet Statistics.   †Including fat and offals.   ‡'000m.
Source (absolute figures): Harry E. Walters and Richard W. Judy, *Soviet Agricultural Output by 1970* in *Soviet and Eastern European Agriculture*, etc., p. 336 (abbreviated). The percentage figures calculated by the author.

this is mainly due to the usual delay in the completion of new processing plants and cheese factories, as the expansion of demand is mainly seen in the direction of whole milk products and of cheese. From 1966 onwards the Soviet Union appeared as a seller of butter-fat at extremely low prices on an over-supplied world market. Thus there is every reason to believe that the Soviet Union should not only manage to meet the domestic demand for milk and milk products in full but that it might contribute to the surplus problem which faces the traditional dairying countries of Europe and Australasia.

The main problem areas for livestock production are, therefore, meat and eggs. With a sustained, even if slow, growth in the supply of fodder grains there is no reason to believe that they should present a very grave problem. Beef and veal output in the Soviet Union, as in most other countries, is closely tied to the dairy herd. As with milk, there is ample room for obtaining more meat from the existing very large herd of cattle, for the average live weight at the time of slaughter is low and the killing-out percentage, i.e. the effective yield of meat per unit of live weight, was only 44.6 per cent in 1962.[2]

The most economic sources of expanded meat supplies are, however, pigs, poultry and, in some areas of the Soviet Union, sheep. Pig-keeping and poultry-keeping lend themselves exceptionally well to factory production. Existing efforts to expand large-scale modern units for both purposes could well bear fruit in substantial and relatively cheap expansion, though the large economies of scale in these industries cannot be realized without careful attention to growing health hazards and without the provision of reliable supporting services.

To sum up, in quantitative terms the rate of growth of agricultural gross production (and even more that of market supplies) in the generally not very favourable period from 1959-61 to 1965-67 was not seriously out of line with the long-term trend of demand for most agricultural products and well ahead of it for some. Grain production, however, still did not provide a really satisfactory margin to cope with the effects of wide fluctuations from year to year; these can only be neutralized by the creation of large buffer stocks to guard against harvest failures such as that of 1963. A further near-record crop in 1968, coupled with a substantial growth in the potato crop in 1967 and 1968, should greatly improve the outlook for the immediate future and assist in the building up of reserve stocks. There is also a need for improving the quantity and quality of fruit and vegetable supplies—and a very large potential demand for imported subtropical and tropical fruit and for the main beverages.

At the same time, the continuation of past trends of growth in the production of certain 'technical' crops and of milk and butter is likely to result in growing export surpluses. It is, therefore, possible to foresee the need for a modification of the policy of indiscriminate expansion at any price which has been pursued, as a rule, since 1953. With a more selective approach, improvements in yields and reductions in the cost of production will become increasingly urgent, though in theory these have, of course, been preached for many years.

2. HOW MUCH AGRICULTURAL INVESTMENT DOES THE SOVIET UNION NEED?

One of the salutary consequences of the disappointment of the more extravagent expectations put in the 'new lands' campaign was the greater emphasis on *land improvement* as a key objective of capital investment in the 1966-70 Five Year Plan.

Without going back to the vast concepts of the Stalin Plan for Remaking Nature, considerable efforts are being set on foot for improving the available land resources and making their use less subject to the vagaries of the seasons. In the south-east this means, above all, increased irrigation as a protection against drought and a condition of improved yields. In the north-west, where the drought hazard is practically non-existent, there is a surfeit of marshy lands which can be made productive through drainage. Acid soils can be improved through liming and land erosion through wind and water must be fought at all times, if the loss of valuable soil is to be prevented or reduced to a minimum. Much of this work needs time and large capital investment which will be amply repaid through higher and, above all, more stable yields, thus reducing the dependence of the country on a niggardly and capricious climate and creating the basis for the successful use of mineral fertilizers on the massive scale permitted by the expansion of the chemical industry.[3]

An essential pre-condition of the modernization of agricultural production is the creation of an adequate rural *infrastructure* which is also needed for improving the quality of life of the rural population. Its key elements are the creation of an adequate network of all-weather roads, the introduction of modern communications and the provision of electricity both for domestic use and for production purposes.

Electricity—in Soviet parlance the electrification of the village—has made quite substantial progress during the last few years. In 1958 almost all State farms but fewer than half the collective farms obtained electricity in one way or another, but by 1965 no fewer than 95 per cent of the latter as well as 99 per cent of the State farms were in this position, and there were very few where electricity was only available for domestic consumption (mainly lighting) and not for use on farm work.[4] (The expansion of rural electricity consumption since the death of Stalin has been fairly rapid. It was only 2,742m. kwh in 1953[5] and as much as 21,000m. in 1965, though this was only a little over 4 per cent of the total output of 506,709m. kwh,[6] and only one half of it was used to power farm machinery.)

In recent years electricity has been increasingly supplied by long-distance cables from large Government-run generating

251

stations, which had been actually forbidden before 1953, when the collective farms and State farms had in consequence to set up their own small and inefficient generating stations.[7] The new policy is undoubtedly more rational and economic in the use of fixed and current resources, but the extension of the grid through long power lines and sub-stations will absorb large capital resources over a long period which will have to be provided by the Government rather than by the collective farms.

The backwardness of the road network and the lack of all-weather roads in the Soviet Union is notorious. With the growth in the size of the public farms this must be a serious obstacle to efficient operation even on intrafarm movements, not to mention transport between the farm and its sources of supply and its markets. With the provision of more lorries improvements in road conditions are becoming increasingly more urgent; whether new roads will be financed by the farms themselves or by the State, they will make growing claims on capital resources.

In addition to land and rural infrastructure, the capital equipment of most public farms will need drastic renewals and additions. This applies in the first place to the *productive buildings,* such as proper storage facilities for fertilizers, crops, etc, and, above all, to the housing of livestock which in Soviet climatic conditions is particularly important and which remains particularly defective in construction, size and quality.

In January 1963 the proportion of livestock in public agriculture kept in so-called main buildings was for *cattle* 81.8 per cent in collective farms and 67.3 per cent in State farms, for *pigs* 83.4 and 72.4 per cent, for *sheep and goats* 70.5 and 61.7 per cent. and for poultry 95.3 and 87.7 per cent respectively.[8] The more unfavourable conditions in the State farms were probably due to the rapid expansion of livestock holdings through the conversion of 'backward' collective farms. In many cases the usefulness of even these 'main' buildings was very limited, for stables had been constructed without overall planning and were largely unsuitable for the mechanization of operations.[9] In January 1962 there were about 41,000 collective farms but 242,000 cowsheds with only 57 cows on average, as well as 170,000 calf sheds, 148,000 piggeries and 91,000 poultry houses.[10] In order to get over the generally uneconomic dispersion of enterprises more is needed than instructions from the centre. Even where buildings are by and large suitable for their purpose, the lack of control over the 'micro-climate' may be a false economy, because it causes high losses through stock mortality and inflated feed costs.[11] Thus large capital investments will have to be made in buildings, partly to ensure suitable accommodation for the growing numbers of productive livestock and partly to permit the mechanization of animal husbandry operations. It is possible to glean an idea of the orders of magnitude involved from the fact that buildings accounted at the beginning of 1966 for 48.5 per cent of total capital assets in public agriculture which was equivalent to some 25,000m. rubles.[12] Only sums amounting to a substantial proportion of this figure can make a real impact on the situation.

Another costly, though probably sound, policy is the concentration of primary processing plants in rural areas. This will both cut down transport costs and improve the utilization of labour in the off-season, but it will need heavy investment in factories and plant.[13] Nor are the existing *machinery* resources adequate for a modern agricultural system either in quantity or in quality. With the exception of the main operations of sowing, ploughing and harvesting of a few staple crops such as grain, sunflower seed and, to a lesser extent, sugar-beet, potatoes and cotton, as well as silage making, agricultural mechanization still has a long way to go, particularly for the auxiliary activities involving the loading and unloading of heavy weights. Inadequate numbers of tractors and harvesters mean loss of time at one of the most critical points in the cropping season; harvesting before maturity or long-drawn out harvesting means everywhere, and particularly in the Russian climate, severe losses in yields.[14] Concentration on wrong types of machinery and insufficient tractor speed wastes valuable time and scarce personnel.[15]

The mechanization of animal husbandry is, as a whole, still in its infancy. In 1965 only 27 per cent of the cows in State and collective farms were machine-milked, and even water was laid on in only half the number of cattle farms and two-thirds of the piggeries, while feeding and manure disposal was almost

253

invariably carried out by hand.[16] The lavish use of manual labour is one of the main causes of the high cost of meat, dairy produce and eggs. Thus a virtually inexhaustible market is assured for the new agricultural machinery factories to be built in the course of the current Five Year Plan.

Finally, the supply of fertilizers, weed-killers and insecticides, of antibiotics on an ever-increasing scale, what the current jargon of Soviet agriculture describes as its 'chemicalization', is putting large additional claims on the chemical industry which can only be met by a massive increase in its capacity.

This short recital of the capital needs of Soviet agriculture reveals large unsatisfied gaps in resources in practically every major branch of activity and type of asset. Sceptics may doubt whether this new investment would be used in the best possible manner, but it is a fact that Soviet agriculture is starved of modern inputs in comparison with advanced Western models. In 1962, the Soviet Union used 62 kg. of mineral fertilizers per hectare of arable land compared with 229 kg. in the United States and employed one 15-h.p. tractor for 90 hectares of sown area compared with 24 hectares in the United States. Similarly, while the energy supply at the elbow of the average agricultural worker in the Soviet Union—in 1965 it was 7.7 horse-power[17]— is one half of that at the disposal of the industrial worker, in America it is claimed to be the other way around.[18] One of the consequences of this discrepancy, or perhaps one of its causes, is the fact that American agriculture, with its somewhat smaller volume, consumes 30,000m. kwh[19] or almost half as much again as Soviet agriculture.

Such comparisons may be too facile to be completely convincing, but in conjunction with the facts of the situation as seen by competent Soviet specialists they add force to the conclusion that the demand for capital investment in Soviet agriculture is very strong and likely to remain so. To the extent that present weaknesses in production, whether in the form of high costs or in that of low yields, can be specifically attributed to the lack of suitable investment goods, there is in general a strong case for supplying the missing tools as soon as possible, though their opportunity cost may sometimes be very high. As agriculture is only one of the claimants on the limited capital resources,

Soviet planners will find plenty of employment in detailed cost-benefit analyses.

Though the needs of increased and more efficient agricultural production are so strong that investment will be mainly geared to directly productive purposes, the continuing low level of public amenities and the virtual absence of a solid infrastructure of modern civilization in wide areas of the rural Soviet Union will require growing attention, if agriculture is to retain adequate resources of skilled manpower. In the State farms, whose staff is classified as workers and employees, the main source of supply for these services is the Government. In the collective farms, their costs have been defrayed out of the 'common consumption fund' fed by allocations from the net income of the farm. The Government is, however, responsible for war pensions, family allowances, industrial injury pensions, health and education and for the major share of the State retirement pensions for which collective farmers have been eligible since 1965.

From the point of view of economic policy the most important aspect of the present situation in Soviet agriculture is the fact that in the foreseeable future there is a very large demand for new capital investment in agriculture. This is proportionately much greater than the need for increased agricultural output, which may soon become much less pressing than the simple extrapolation of past experience would suggest at first sight. However, in order to meet this relatively modest additional demand—and, above all, in order to meet it more efficiently—a very large back-log of demand for capital investment of all kinds will have to be worked off. The neglect of generations cannot be made good in a single spurt, and agricultural investment will remain for some time one of the biggest items in the capital budget of the Soviet Union.

The most critical investment problems are partly physical and partly financial and the sources of finance and the terms on which it is made available to agriculture strongly affect agricultural policies. Up to 1957 the Government provided, in addition to the fixed resources of the State farms and subsidiary public farm enterprises, through the MTS a large part of the machinery resources of the collective farms; other investment by the latter was supplied through allocations to the indivisible fund which took precedence over the earnings of the collective farmers for their work.

The transfer of the MTS to the collective farms had two important financial effects: the Government recovered part of its earlier outlay through the repayment by the collective farms of 1,800m. rubles for the tractors and machinery taken over by them, and the collective farms had to shoulder the financial responsibility for further investment in agricultural plant.[20] On the other hand, the Government took over the financing of investment on collective farms converted into State farms. These changes contributed to the decline in Government investment in agriculture in 1958-59, while investment allocations by the collective farms reached a peak in 1959 which they did not again surpass until the end of the Khrushchev era.

During the period of the current Five Year Plan, Government investment in agriculture should double to reach 41,000m. rubles; though its share in total agricultural investment during 1966-70 will rise to about 60 per cent, the collective farms will also increase their own investment by 50 per cent to 30,000m. rubles. This amount has to be viewed against the background of a total money income of the collective farms from all sources of 19,900m. rubles in 1965 of which 10,400m. came from crops, 8,600m. from animal products and the relatively small balance from subsidiary enterprises.[21] The current Plan envisages an increase in agricultural gross production on average by 25 per cent over the previous five years. As gross production in 1965 was about 7 per cent higher than the average for 1961-65 (Appendix I), this target is about 17 per cent higher than the actual level in 1965. Assuming for the sake of simplicity that the collective farms will exactly conform to the average, they would increase their money income at unchanged prices and relative volumes of operations to slightly less than 23,000m. rubles per year, out of which they are intended to invest no less than 6,000m. or slightly more than one-quarter.

The broad categories employed by Soviet economists in their analysis of agricultural production are exceedingly simple. They start with gross production which includes not only the value of final agricultural output but also that of produce such as seed and feed used in the production of final output. If such items as well as other current inputs and even depreciation are deducted from gross production, the balance represents agricultural gross income. This consists of two parts—payments for labour and agricultural net income which is the amount available for investment, allocation to insurance funds and tax payments.[22]

With the overdue change-over in the collective farms to the payment of fixed money wages, the allocation to investment and insurance reserves, etc will become the residuary legatee whose share will depend on the excess of receipts over other outlays. As the agricultural tax has also been changed into a tax on net income, it may be said that the agricultural net income will for the first time become a genuine balancing figure. However, if this must be large enough to permit the annual investment of 6,000m. rubles, producer prices of agricultural produce must be set high enough to make this possible, and in view of the dominant position of the Government as a purchaser of collective farm output, this involves a correspondingly high level of procurement prices.

After forty years of changes which, however brutal and inefficient they may have been, have completely altered the social and economic structure of the Soviet village, the wheel seems to have come full circle. The clash between the Soviet power and the peasants was caused by the incompatibility between the growing needs of the Government for agricultural produce and the lack of additional industrial goods to give the peasants in exchange for bigger output. The forced contribution of the peasants to the cost of industrialization took the form of depressed producer prices and ultimately of the virtual confiscation of the 'marketable surplus' of Soviet agriculture. The Soviet power has now accepted the need for making good much of the arrears in agricultural investment which have accumulated over decades, but although the larger part of this investment will come directly from public funds, a substantial proportion will be defrayed by the collective farms—though no longer at the expense of the living standards of the peasants but at that of the economy as a whole in the form of high producer prices.

3. HOW HIGH ARE SOVIET PRICES AND COSTS?

Soviet Producer Prices in Perspective

Pricing methods and price levels of agricultural products in the Soviet Union, as in most other developed countries, are much more artificial and politically motivated than prices in most other sectors of the economy. Even the so-called unified producer prices introduced in 1958 were by no means so simple as this description implies; one of the main reasons why Khrushchev's successors returned to the two-tier prices abandoned in 1958 was, indeed, the uncertainty and the arbitrary character of the new system introduced at that time.

A comparison of the average procurement prices of various agricultural commodities in the last year of the Stalin era and in 1963, as shown in Table 49 (p. 201), reveals in the first place the enormous increases in the prices of most products during this period. The figures can, however, also be used to show the price relationships for these products and the changes in these relationships over time. A convenient but not invariably suitable way of presenting price relationships is the use of the wheat price as a basis and this method is used in the next table.

TABLE 61
Soviet average procurement prices 1952 and 1963 in terms of wheat

	1952	1963
Wheat	100	100
Maize	56	101
Raw cotton	3,287	507
Sugar-beet	108	38
Sunflower seed	198	238
Potatoes	48	94
Beef cattle*	209	1,057
Pigs*	693	1,296
Milk	260	161
Eggs†	3,487	1,574

*Live weight. †Calculated at 17 eggs = 1 kg.
Source: Calculated from Table 49.

Between 1952 and 1963 the prices of raw cotton, sugar-beet, milk and eggs had fallen relatively to the wheat price and those of maize, potatoes, sunflower seed and meat had risen; for cotton, beef, cattle and eggs (as well as for wool) the change was dramatic.

It is tempting to supplement such meagre deductions with international comparisions of producer prices in the Soviet Union and Western countries, though these are subject to very severe qualifications. The first difficulty is the conversion of prices from one currency to another. This is never quite satisfactory, but when the currencies are convertible and the products are in practice exchanged it is possible to attach some practical meaning to the results.

However, if one of the currencies is formally inconvertible and belongs to a country whose social system differs radically from that of the other countries and where prices (and costs) are determined on different principles from those of the rest, the same nominal sum may mean something very different. This affects not only the absolute price comparisons (of the type 1 ton of wheat = x rubles, 1 ruble = y $, 1 ton of wheat = x.y $) but also the price ratios between different products. This introduces the second difficulty, that the relative prices of different commodities vary in any case widely from one country to another and none of them can be reasonably regarded as a standard capable of general application to which all others ought to conform.

The following table presents price comparisons on a wheat basis for selected countries of Western Europe and the United States in a recent year; for commodities whose prices vary sharply from year to year, such as potatoes, the use of a single year has, of course, serious disadvantages and in the year in question potatoes were unusually dear in most European countries but cheap in the United Kingdom.

In this table countries have been arranged in rising order of the producer price of wheat and the result suggests that the wheat price may not, in fact, be a particularly sound basis for such a comparison, because its level varies much more than the prices of some other commodities, especially meat but also sugar-beet and milk; hence there is a general tendency for the price ratio of many commodities to wheat to fall with rising national wheat prices. This makes it impossible to determine any single set of price relationships which fits countries with basically different agricultural and economic conditions.

Thus the European Commission of the EEC employs as a rule

TABLE 62

Relative producer prices in selected countries

	USA (1965)	UK (1965–6)	France (1966)	Belgium (1966)	Netherlands (1965)	Germany (1965–6)	Italy (1965–6)
Wheat—$ per ton	49	68	86	95	98	105	111
			(Wheat=100)				
Sugar-beet	31	26	15	20	18	18	17
Potatoes	114	58	56	60	52–59	48	70
Beef cattle*	896	730	728	688	617	577	620
Pigs*	926	730	763–822	664	545	645	600
Sheep*	1,020	728	939	455	414	510	661
Milk	191	143	94	87	94	96	92–100
Eggs†	—	1,075	715	764	527	735	677

* The price ratios relate to live weight; where only dead weight figures were available, the following killing-out percentages were used: cattle 55%, pigs 73%, sheep 50%.   † 17 eggs=1 kg.

Sources: FAO Production Year Book 1966. UK: 1967 Price Review White Paper (Cmnd. 3229). EEC: Agrarstatistik 1967, No. 4.

of thumb for the appropriate price relationship between wheat, milk and beef the ratio of 1:1:7, i.e. that the price per ton of wheat should roughly equal the price per ton of milk and that the price per ton of beef cattle (live weight) should be at least seven times that of a ton of milk. Simple inspection of the situation in the EEC countries suggests that this fits the actual state of affairs in these countries well enough, though this is hardly a reason to attribute to it any normative value. However, any attempt to apply this set of ratios to the United States or to Britain would involve either a drastic reduction in milk prices (and in the United States a sharp fall in beef prices) or an enormous rise in the wheat price. In the Anglo-Saxon countries the price ratio between milk and beef is roughly the same at 1:5; this is much higher than the EEC ratio of 1:7, and it would be no less futile to claim that either the one or the other of these ratios is 'correct' than to determine whether the absolute level of the wheat price in the USA is more or less justified that that in the EEC.

The absolute level of the wheat price paid to producers in the Soviet Union in 1963 was about 84 US dollars at the official parity between the dollar and the ruble. This was half way between the actual US price and the highest price paid in any EEC country about two years later; at that time the Soviet prices had been increased by varying amounts in different parts of the country and for various types of procurement and may have been, on average, not very different from the EEC target price. If it is assumed that the price increases for most animal products were proportionately similar to those for wheat it may be guessed that the price relationships did not change very much, though this will have to await confirmation in the form of official calculations of average procurement prices.

At least in 1963 potatoes and sugar-beet were priced in the Soviet Union in a broadly similar manner as in the United States in relation to wheat, milk was relatively somewhat cheaper and beef a little—and pig meat considerably—dearer. The milk: beef price ratio was about $1:6\frac{1}{2}$—nearer to the EEC position than to that in the United States, but not outrageously out of line with that in other countries. In general, however, most Soviet price ratios would seem to be broadly comparable to those in the United States where wheat is cheap rather than to those of the EEC countries where wheat is dear.

As the Soviet producer price of wheat, expressed in US dollars at the official rate of exchange, is obviously much higher than in the United States it is an easy matter to demonstrate that on this assumption agricultural producer prices in the Soviet Union, and particularly the prices of animal products, are extremely and unreasonably high.[23] It will be seen that the efficiency of livestock production in the Soviet Union is, indeed, low but price comparisons between the Soviet Union and the United States in absolute terms involve an answer to the question whether one ruble is, in fact, worth more than a US dollar and in this general form the question is completely unanswerable. Comparisons of price relationships lead to the just as insoluble problem whether, e.g. the 'Anglo-Saxon' milk: beef price ratio of $1:5$ is preferable to the 'Continental' European ratio of $1:7$ or whether there is any economic sense in assuming that either the one or the other can be applied to Soviet conditions without a detailed analysis of the facts. The only fair and honest conclusion is, therefore, that agricultural price comparisons between countries with widely differing conditions are in the nature of things inconclusive.*

Comparative Costs of Production

Reasonably reliable estimates of agricultural costs of production for individual commodities are in all countries much more difficult to obtain than accurate price statistics. They are as a rule only available for farm samples which may not be fully representative and it is far from easy to ensure that all costs have been analysed according to the same accounting techniques, particularly the very important overhead costs and joint costs. In the Soviet Union the strength of central bureaucratic control, whatever its defects in other respects, may at least have caused a fair degree of uniformity in the construction and presentation of cost accounts. Figures of average total costs of production for the main agricultural commodities have been published in recent years in the form of All-Union and regional averages for State farms and collective farms separately, though without statistical indications of the dispersion of actual results around the mean in a form which would make it possible to determine their significance.

Unfortunately the application of these figures is severely restricted in two respects. The changes which have occurred in recent years between State farms and collective farms as a result of the conversion policy in operation from 1957 to 1962 or even later make it impossible to compare cost trends even for the few years for which average costs are available. If State

* The pitfalls of such comparisons may be illustrated in extreme form by a curious error in which Professor D. Gale Johnson, one of the most respected American agricultural economists, became involved. He compared the milk: beef price ratio in the United States and the Soviet Union in the early part of the Khrushchev era and was puzzled by his finding that in the Soviet Union 'beef prices are about six times as high compared with the United States as milk prices'. As on his own figures the Soviet milk: beef price ratio was at the time about 1:3 and one of the highest in the world, this would, indeed, have been puzzling, for in the United States it was at the time about 1:5. In fact, Professor Gale had overstated the American milk price tenfold by calculating it as $88.8 per 100 kg. which gave the nonsensical milk: beef price ratio of 2:1. However, he was apparently so convinced of the validity of the American price ratios emerging from this exercise that he concluded: 'In fact (!), the low milk output per cow may be in part a consequence of the low price of milk compared with beef.'[24]

farm costs have tended to increase, to what extent has this been due to the absorption of high-cost collective farms? Conversely, how far have changes in collective farm costs been caused by the presence in one year, and the absence in another, of groups whose cost levels were substantially different from the average?

Even more intractable are the difficulties connected with the valuation of labour in the collective farms. Labour is the single most important cost element, which accounted for about 40 per cent of total gross costs in public agriculture in 1962.[25] For most of the period covered by the available statistics this was treated in the collective farms as a residual item, the balance remaining after the deduction of all other charges, including investment allocations, from gross income. The remuneration of labour depended, therefore, not only on the level of these other charges but also on that of gross returns which reflected, above all, the procurement prices paid by the Government for collective farm produce. Increases in producer prices directly increased the cost of production in the collective farms and any attempt to measure the effect of price increases on the balance between prices and costs *in total* thus involves circular reasoning. Given the revolution in producer prices after 1953, comparisons of the trend of production costs would thus be self-defeating even if they could be made over a sufficiently long period. This difficulty is not so great for individual commodities, particularly during the years when the remuneration of labour depended primarily on the number of labour days, whose value was calculated in relation to the gross income (less investment) of the collective farm as a whole. To the extent that producer prices were set for individual commodities with a view to guiding output into certain directions, it can be fairly argued that this could be achieved within the framework of the existing payment structure for collective farm labour.

As for international cost comparisons in terms of value the difficulties are even greater than those encountered for price comparisons, partly because the available material is everywhere subject to qualifications relating to the specific social and economic environment in which the costs are incurred. Such comparisons are most fruitful in terms of physical and technical coefficients, such as output per acre, labour usage per unit of product, feed conversion rates etc, although even then the disregard of their limitations invalidates the results for any purpose except that of propaganda. Nevertheless, as far as the Soviet Union is concerned, most comparisons of this kind, whether carried out by Soviet specialists or by Western critics of Soviet agriculture, support the conclusion that technical efficiency in most branches of Soviet agriculture is well below that of more advanced—and generally climatically more favoured—Western countries.

The Profitability of Agricultural Enterprises in the Soviet Union

One field in which the comparison of production costs and prices is practically important and, therefore, necessary is that between the costs incurred and the prices received by agricultural producers for the same commodity, which determines the profitability of producing it. As the payment of low producer prices has been historically the main form of exploitation of the peasants by the Soviet régime, such comparisons are of special interest.

TABLE 63

'Actual' collective farm costs, producer prices and profitability, 1962–64
(rubles per ton, eggs, rubles per '000)

	'Actual' costs		Average price	Profit or loss (−)	
	(1962)	(1964)	(1963)	(1962/3)	(1963/4)
Wheat	37*	44*	75.6	38.6*	31.6*
Raw cotton	224	281	383	159	102
Sugar-beet	16	17	28.7	12.7	11.7
Sunflower seed	30	30	181	151	151
Potatoes	38	35	71	33	36
Cattle†	834	927	799	− 35	−128
Pigs†	1,146	1,250	980	−166	−270
Milk	129	151	121.8	− 7.2	− 29.2
Eggs	—	82	70	—	− 12
Wool	2,504	2,939	3,786.7	1,282.7	847.7

*The costs relate to all cereals except maize and the comparison is, therefore, even more approximate than for the other products.

† Prices are for live weight, costs for weight increase.

Sources: Costs: Narkhoz 1962, pp. 338f. Narkhoz 1964, pp. 396f.

Prices: Table 49.

Profitability: Calculated from costs and prices.

The costs incurred by the collective farms are published on the basis of two assumptions; one of these values labour at the rates (in money and in kind) actually paid for it during the year in question, while the other uses the rates and norms applicable to State farm workers as the basis of valuation. Despite the element of circular reasoning involved in the calculation of 'actual' collective farm costs prior to the introduction of specific money wages, this is obviously the only realistic basis for an assessment of the profitability of individual enterprises. In table 63 above, costs in 1962 and 1964 are compared with average prices in 1963, partly because of the limitations of the data and partly because of the unusual crop situation in 1963. The range of profits or losses reflects the cost changes between 1962 and 1964.

These results are confirmed from a different angle, and amplified by the inclusion of a number of additional enterprises, by statistics of the net income per thousand rubles of expenses relating to public agriculture as a whole in 1962:

TABLE 64

Net income (in rubles) per '000 rubles expenditure, 1962

Vegetable products		Liverstock products	
All grain	+ 440	Cattle	−489
Raw cotton	+2,696	Pigs	−314
Sugar-beet	+ 368	Sheep and goats	+300
Sunflower seed	+4,420	Poultry	−205
Potatoes	+ 722	Milk	−220
Vegetables	+ 702	Eggs	−153
Fruit and berries	+ 994	Wool	+366
Grapes and wine	+1,735		
Silage	− 321		
Roots	− 437		
Natural pasture	− 257		

Source: Y. Lemeshev (ed.) *Ekonomicheskoie Obosnovanie*, etc., pp. 46, 117.

The position towards the end of the Khrushchev era was thus comparatively clear-cut. The average procurement prices of most vegetable products, including most or all of those grown for sale, were well in excess of their average costs in collective farms, including the cost of labour at the rates actually paid to collective farm members. The net margins were high for the main staple crops (grain, potatoes, sugar-beet) and enormous for cotton, sunflower seed, grapes and fruit. On the other hand, of all livestock products only sheep seem to have been a paying proposition both in respect of mutton and of wool, while all other animal products and the fodder crops raised for them were produced at substantial losses, though the way in which the loss on fodder crops can have been calculated is somewhat intriguing.

This was the starting point for the further price increases made by the new leadership in 1965. They were too complicated to enable any outsider to assess their average effects and still less to calculate the position in different parts of the country. As for cereals, their main purpose was obviously to encourage higher production in the 'non-black-earth zone' regions outside the traditional producing areas, mainly because they are virtually free from drought hazards. Their production costs, however, are more than double those of the northern Caucasus and the southern Ukraine, the low-cost regions, and up to double the national average. In 1964, the collective farms of the southern Ukraine produced grain (excluding maize) at an average cost of 31 rubles per ton, but in central Russia and the upper Volga the cost was 70-75 rubles, in Lithuania and Latvia 88 rubles and in White Russia 85 rubles.[26] (On the State farms, both extremes were even wider, for while the State farms of the northern Caucasus and the southern Ukraine—but not those of Khazakshan—produced grain considerably more cheaply than the collective farms of the same areas, the opposite was the case in the 'non-black-earth zone'.[27])

As for livestock products, the position was much simpler, because they were apparently generally, though perhaps not invariably, unprofitable and the aim was to cover the handicap of procurement prices which did not even cover the cost of production. This situation may be illustrated in somewhat greater detail for milk. In 1964, the Government acquired 31.4m. tons of milk (including the milk equivalent of milk products); 41 per cent or 12.9m. tons came from the State farms, etc, and 18.5m. from other sources, including 2.5m. from the subsidiary private holdings and 16m. tons from the collective farms.[28] The average procurement price paid to collective farms and to private holdings was about 127 rubles per ton and the total revenue of collective farms and private holdings from milk sales to the Government was 2,344m. rubles.[29] As the average cost of production in the collective farms in 1964 was 151 rubles per ton, their loss on milk sales to the Government in 1964 amounted to about 384m. rubles; the losses on milk sales incurred by the State farms, etc, in the same year are quoted at 679m. rubles.[30] The losses of public agriculture on its milk sales to the Government thus exceeded 1,000m. rubles and it was small comfort to State farm directors and collective farm chairmen that the milk processing industry showed record profits of 1,389m. rubles in the same year.[31]

The changes in the producer prices of milk at the beginning of 1965 included increases in basic milk prices and in the number of price zones and adjustments in butter-fat and hygienic quality standards to the benefit of the producers.[32] In the Baltic republics, with their special aptitude for milk production, the basic price was raised to 160 rubles per ton; as the cost of production in this region in 1964 had averaged 123 rubles in collective farms and 147 rubles in State farms, this was enough to ensure on average profitable sales. But in the Moscow area, too, where the new price was 190 rubles, milk production was now generally economic in both State and collective farms, quite apart from the opportunity for selling directly in the liquid market. There returns of 254 rubles could be obtained which left a clear profit of 21 rubles even after allowing in full for the extra costs incurred.[33]

It may be concluded that, by and large, the substantial price increases of 1965 brought all Government procurement prices into line with actual costs of production for those commodities which had been unprofitable before, either generally or in some parts of the country. The Government thus seems to have accepted the fact that in the short run this was the only possible way of stimulating supplies and liquidating the awkward consumer shortages of livestock products which were particularly noticeable in the wake of the 1963 crop failure and its effect on animal husbandry in the following year. In the long run, however, there is undoubtedly an urgent need for raising the efficiency of animal husbandry in the Soviet Union and there is good reason to believe that there is ample opportunity for such a process.

The causes of the present unfavourable position are deep-rooted. They go back at least to the drop in livestock numbers during mass collectivization and are connected both with the still high share of the tiny subsidiary economy of the peasants in the supply of some products and the inefficiency of animal husbandry in the large public farm enterprises.

The two main cost elements of livestock production are feed and labour; in 1962, feed accounted for 44 per cent and labour for over 35 per cent of the total in public agriculture.[34] No radical improvement in the situation can be expected as long as these two cost factors remain at their present levels. Ignoring the structural consequences of the existence of very large numbers of single-animal holdings, on the one hand, and huge concentrations of stock in State and collective farms, on the other hand, there are three outstanding economic weaknesses in Soviet animal husbandry: the insufficiency of fodder, the low yields and consequent inflated numbers of productive livestock and the excessive use of manual labour.

Even after the excellent harvest of 1964, the total availability of feed in 1965 was only marginally greater than it had been in 1962, when it was known to have been well beneath requirements. As there were 5m. more cattle, some form of balance had to be restored by a sharp cut-back in pigs and poultry, which are more efficient converters of feed into meat than cattle but need more concentrates. The position was made even worse by the lack of high-protein feed which led to overfeeding in quantity in order to make up for inadequate quality: the average protein content of a feed unit* was only 80 gr. which is not even enough for sheep and cattle and far too low for pigs and poultry.[35]

The low proportion of concentrates in Soviet feed supplies is, above all, due to the shortage of feed grains. This is particularly important in a country where grassland is at present not the cheapest form of feed for cattle and sheep. In most temperate climates it is regarded as sound policy to base the cattle economy in the first place on grazing and the use of conserved grass in the form of hay and silage, and to supply concentrates (grains,

* The Soviet feed unit is based on the feeding value of 1 kg. of oats.

oil cakes, etc) only if and when it is necessary to supplement grass, silage and other bulk fodder.

In the Soviet Union, feed grain is, on the contrary, one of the cheapest and perhaps actually the cheapest source of energy for feeding animals, at least in public agriculture. In 1962, the cost of a ton of feed units in the form of grain was 33.6 rubles in the collective and State farms; this compared with 37.2 rubles for pasture, 36.4 to 43.3 rubles for hay, 43.3 rubles for silage, 77.3 rubles for sugar-beet and as much as 153.3 rubles for potatoes.[36] With ample supplies of feed grain, it would obviously be madness to feed potatoes and sugar-beet to stock, and the proportions of grains, grass and silage might also come under critical review. In harsh fact, the shortage of feed grain has forced Soviet animal husbandry into dependence on other and much more expensive feeding stuffs, such as potatoes, which is one of the main causes of its high costs.* Neither the size nor the value of the potato crop in relation to grain would be justified, if there had been enough maize and barley for pig-feeding. (The gross value of the production of potatoes in 1958 was 6,934m. rubles compared with 7,953m. for all grain; it constituted more than a quarter of the total gross value of all crops. Though by 1962 this proportion had declined, it still was one-fifth of the value of all vegetable products.[38]) One of the keys to a reduction in the cost of livestock products, except wool and mutton, is a substantial improvement in the availability of feed grains.

Low feeding, and particularly inadequate protein feeding, is directly reflected in poor yields. Just as low-quality feed has to be used in excessive quantity, poor-yielding beasts are kept in excessive numbers in order to obtain output. Better feed sup-

* The late Naum Jasny argued that 'it can be assumed for the capitalist countries that farmers will grow (or purchase) feeds which will cost them, bearing quality in mind, less than grain and that there will be no point in their producing (or purchasing) and using feeds more expensive than grain. At the existing prices the supply of grain is unlimited in the capitalist countries'.[37] Even if the last statement is accepted, the feed grain shortage in the Soviet Union may be a good polemical stick for castigating Soviet shortcomings, but this should not obscure the fact that a country in which concentrates are the cheapest and not the dearest feeding stuff faces more difficult economic problems in livestock production than countries able to draw on a variety of cheaper feed. On the other hand, the current cost relationships between various types of feed in Russia should not necessarily be regarded as an unalterable fact of nature.

plies would very quickly improve productivity, though they would not cure the structural weaknesses of the size distribution of livestock.

The extent of the yield problem may be indicated by a few key figures. The milk yield per cow for the whole of agriculture in the record year 1966 was 1,888 kg.[39] or less than half the average yield of the best dairying countries; even in the large herds of public agriculture it was only 2,021 kg. (In 1968 it is claimed to have reached 2,232 kg., though figures for single years are subject to wide fluctuations.) The live weight of cattle at slaughter in 1962 was 248 kg. compared with 400 kg. or more in Britain and the United States at twelve to eighteen months; pigs reach 100 kg. in the United States in five or six months, but in the Soviet Union their slaughter weight averaged 78 kg.[40] In the Soviet Union many State and collective farms produce only seven or eight piglets per sow per year[41]—in many advanced countries this would be regarded as too few for a single litter. In 1968 the average number of eggs per laying hen was 152 (in public agriculture)[42] compared with 187 in Britain and 210 in the United States per bird in the flock at the beginning of the year.[43]

The ratio between the feed needed for maintenance and that available for production varies according to the type of animal, but the very low yields prevalent in the Soviet Union invariably have the effect that a smaller proportion of the total feed intake is converted into livestock products than in more efficient systems. This means that, on a national level, the total herd or flock is far too large and consumes too much of the available feed for its own maintenance. It is too early to tell whether the drop in the number of cattle and the much sharper decline in the number of pigs in 1967 and 1968, which was a period affected by foot-and-mouth disease, indicates a deliberate policy on the part of the authorities[44] but there is little doubt that more livestock products could be produced from smaller herds and flocks with a substantial improvement in the utilization of feed.

There is no less room for improved efficiency of labour, though most of this will require large capital investment in plant and buildings. In 1962, milk production was by far the greatest single user of labour: with 15.2 per cent of the total it absorbed more labour than the whole of grain production, including maize. Beef cattle took 8.2 per cent, pigs 5.6 per cent and sheep 4 per cent of total labour resources. Animal husbandry as a whole needed 34.6 per cent, vegetable products 41.6 per cent and 'indirect' labour 23.8 per cent of the total.[45] According to Soviet calculations, on the average collective farm in 1962 one dairy-maid looked after thirteen cows and one worker was responsible for eighty-eight pigs.[46] In terms of man days per 100 kg. of product, the direct labour requirement (in 1964) was 2.70 for milk on collective farms and 1.76 in State farms, for beef 13.6 and 8.4 respectively, for pigs 13.8 and 5.6, and for sheep 10.3 and 6.9.[47] (The figures for meat relate to 100 kg. weight increase.)

Increase in productivity is the key to better remuneration of labour combined with stable or falling cost of production. Official Soviet sources tend to emphasize the share of insufficient mechanization in the present unsatisfactory position; this attitude ignores the structural weaknesses of animal husbandry in the Soviet Union, but despite its onesidedness it is far from baseless. The high cost of labour is certainly connected with the widespread prevalence of manual work in feeding, including watering, livestock, in milking and in manure disposal. There is an almost unlimited need for capital investment in infrastructure (water and electricity supply), buildings suitable for mechanization and plant (e.g. milking machines) in order to create favourable conditions for a cut in the labour component of meat, milk and eggs. At the same time there is just as great a need for the improvement in the size structure of livestock holdings and a consequent reduction in the high cost of administration in public agriculture.

4. ARE SOVIET FARMS TOO LARGE?

1962 was the high-water mark of the extension in size of State and collective farms. Public agriculture consisted, in the main, of 8,570 State farms and 39,700 collective farms. The average sown area of the former was about 10,100 hectares and that of the latter 2,800 hectares; including meadows and permanent grassland, the average State farm measured over 28,000 and the average collective farm over 6,000 hectares. Such enormous land resources were more than matched by the size of the labour force, for the State farms employed on average a staff of 825—considerably more in specialized cotton and cereal farms—and the average collective farm embraced 404 households, and about 500 people actively participated in the public enterprise.[48] In extreme cases there were State farms with an area exceeding 70,000 hectares of agricultural land or 700 square km. and which included 50 to 70 'populated points' separated by distances of 60 to 70 km.[49]

The livestock holdings of these huge enterprises were also very substantial, with the average State farm owning 2,447 head of cattle, including 863 cows, 1,956 pigs and 4,798 sheep and goats. Collective farm herds and flocks were smaller, but even there the average was 954 head of cattle, with 336 cows, 787 pigs and 1,667 sheep and goats.[48]

These figures do not give an adequate idea of the size structure of public farms, but they demonstrate the heavy preponderance of very large units with nominally huge labour forces such as are very rarely found in most advanced countries. In this respect Soviet agriculture is more comparable to the latifundia in parts of southern Europe or South America; the subsidiary private economy of the Soviet farm population, with its diminutive garden plots and livestock holdings, may also be compared to the small plots of many agricultural labourers on such large estates.

Historically, the creation of large-scale farming in the Soviet Union had nothing to do with reducing costs of production or obtaining the highest possible income for the peasants. The dominant motives were the need of the towns and those of industry for more agricultural produce and the difficulties of organized 'procurement' from some 25m. peasant households. Collectivization consisted in the agglomeration of the tiny peasant farms into still relatively small collective farms, while the experience of the 1930s with huge State farms was so forbidding that their numbers remained very small. From 1950 onwards the authorities started a deliberate and sustained campaign to reduce the numbers and increase the size of collective farms through amalgamations. This was supplemented from

1957 onwards by the conversion of collective farms into State farms; as the new lands campaign led to the creation of many new and very large State farms in the development areas, the result was the pattern of very large farms which reached its probably highest point in 1962, when State farms were on average three or four times larger than collective farms.

Both types of public farms are regarded as financial and economic units. This is particularly important for the collective farms where until recently even the labour income of the 'members' was based on the value of the labour day which was calculated for the collective farm as a whole. The introduction of compulsory money wages payable at more or less regular intervals has changed this situation, but the collective farm is an economic unit in much more than an accounting sense and is financially responsible for the bulk of its capital investment.

The conversion of collective farms into State farms was at one stage largely a way of avoiding the rigidities of this system, for the fixed capital of the State farms was always provided by the Government and wages based on certain norms were paid irrespective of their financial results. In these respects the two main forms of public agriculture represented opposite extremes of financial organization, but the introduction of firm wages in the collective farms and the growing practice of using part of the surplus of the individual State farm for bonus payments, etc, have narrowed the gap between them.

Neither historically nor in their current operations are the public farms fully centralized enterprises. In practice they include a considerable measure of differentiation of functions, with the broad managerial direction and certain service tasks organized at whole-farm level and the organization of production delegated to smaller units.

Administration in Public Farms
In economic terms, important differences between the two main types continue. The collective farm management plays a significant part in marketing a proportion of the output and in allocating gross income between investment and consumption; in the State farms, on the other hand, investment is mainly, and

the level of wages and salaries largely, independent of profit or loss, though they are supposed to be financially self-supporting. The functions common to both types which are performed centrally comprise administrative and technical direction, including particularly the detailed setting of production targets in line with the Government procurement plans as well as certain central services such as parts of the accounts, veterinary assistance, plant repairs, building work and sometimes even electricity supplies: 'For the administration of the collective farm and direction of the State farm is reserved the organization of production plans, the supply of the essential means of production to the brigades and other subdivisions, the organization of the transfer of produce to the Government, the definition of the order in which production takes place, the organization of the accounts and the essential control over the work of the sub-divisions.'[50] In addition, State farms and collective farms—particularly the latter—also perform certain social functions (provision of nurseries, libraries, etc.) which might in other countries be regarded more as the sphere of local government.

Despite Khrushchev's tirades against chairborne specialists and ministerial bureaucrats, the proportion of educationally qualified agricultural experts in the central and regional administration remains very high; at the end of 1964 there were 178,000 compared with 457,000 specialists with higher and secondary education who actually worked on farms. In every bureaucracy the centre is the seat of power and the fountain of promotion, and the tendency for the best qualified staff to congregate there remains obviously very strong. Nevertheless, the corps of reasonably well educated officers in State and collective farms is numerically very large.

The following table shows the balance of managerial and technical staff within the two main forms of public agriculture and the proportion of educationally qualified personnel in each group. The figures exclude the numerous office staff, but they give a broad view of the managerial and technical structure of the average public farm enterprise. It may be assumed that the majority of the professionally qualified staff generally operates at the centre of the farm and that the local managers are mainly the brigadiers and heads of livestock units, though on large farms they may have the advice of agronomists or livestock experts working at the brigade level.

TABLE 65

Managerial and technical staff of public farms on April 1, 1965

| | '000 persons | | % with higher and secondary education | |
	Collective farms	State farms	Collective farms	State farms
Chairmen or Directors	37.4	10.5	67.3	91.8
Deputy chairmen	13.3	—	40.6	—
Divisional managers	—	33.5	—	49.8
Chief specialists	—	38.9	—	94.8
Agronomists	47.4	21.7	94.9	89.6
Livestock experts	38.7	19.5	90.6	82.2
Veterinary staff	45.9	28.8	48.6	65.4
Engineers, technicians	43.6	31.9	40.7	38.8
Brigadiers	181.0	70.0	10.1	22.3
Livestock managers	89.0	46.3	10.8	17.7
Total	496.3	301.1	—	—

Source: Narkhoz 1964, pp. 423ff.

An over-simplified picture of the top management structure of public farms may be constructed from this information. In the *collective farms* the chairman is assisted by a deputy on about one farm in three; he has at his disposal one livestock expert and one (or exceptionally two) agronomists, veterinary experts and engineers. This group of about six persons, plus the chief accountant, forms the directing personnel of the whole farm. The direct supervision of the productive process is in the hands of a slightly larger group of perhaps five brigadiers and three livestock farm managers. Agronomists and livestock experts tend to have the highest proportion of educationally qualified staff and two out of three chairmen have at least had secondary education. At the other extreme, only one in ten of the local managers have had more than elementary education.

The *State farms* have a much broader top managerial and technical superstructure. They may be three times larger than the collective farms, but every director disposes on average not only of three divisional managers but also of three to four chief specialists, each of whom has a staff of several (two or three) skilled, though not necessarily highly educated, experts. In contrast to the collective farms, the central group of top managers and specialists in the State farms numbers about eighteen and is almost twice as numerous as the local brigade and livestock managers. In numerical terms there is thus no evidence whatever of economies of scale in the employment of managerial and technical State farm staff. Educationally, there is little difference between the two types for the common run of technical experts, but a much higher proportion of the State farm directors than of collective farm chairmen is educationally qualified; in addition, the State farms have their high-powered chief specialists for whom there is no equivalent in the collective farms. The local managers of the State farms are also significantly better educated than those of the collective farms.

The overall proportion of man-power not directly engaged in productive operations is very similar in both types of public farms. In 1959-61, the total labour used for cereals on public farms averaged 8.76 man-days per hectare; of this, 'direct' labour amounted to 6.60 man-days and 'indirect' labour of all kinds absorbed 2.16 man-days or almost exactly one-quarter,[51] but a good deal depends on the definition of these categories.

Though it is impossible to allocate the educationally qualified staff between central and local management, the low level of education of the local managers suggests that most of the better qualified staff are concentrated at the central farm level. The very high complement of professional staff in the administration of public agriculture suggests that at least one of the reasons for the progressive reduction in the number and increasing size of public farms, the shortage of managerial staff, is no longer valid. It may, indeed, indicate the opposite, though this is difficult to prove with the available evidence.

The need for such very large specialist staffs at farm level may be partly due to the size of the farms themselves; one specific reason is the amount of work involved in the detailed spelling out of production targets given to the individual farm (or formulated by it on the basis of the Government procurement plans) into production tasks for the brigades or other sub-divisions.[52] Another reason is the weakness or lack of supporting institutions both in the supply of the necessary services and in the marketing of produce.

The network of auxiliary industries and services in the Soviet

Union is undoubtedly very backward for a system in which the greatest possible use of modern equipment is an article of faith, even though the actual degree of mechanization is still patchy. In the agriculture of most advanced countries, and particularly that of the United States, the production unit 'is supported by a vast array of large-scale units, some in the public sector providing research, adult education, transportation and communication facilities; some in the co-operative, semi-public sector providing input supply, marketing, credit; and some in the private sector supplying essentially similar services to those in the co-operative sector and in competition with that sector.'[53]

This statement may convey a somewhat idyllic view of the relations between the large-scale units of private enterprise and the individual farmer, both in the supply of agricultural inputs and in the purchase of agricultural output, but, the disparity of power between agricultural producers and the business firms providing such services apart, the services themselves are also needed on Soviet farms and the under-development of outside services justifies, at least for the time being, their provision at farm level. A farm advisory service, veterinary assistance, plant repairs and construction jobs, road transport or even electricity supplies which are supplied on the level of large public farms may be less efficient than the specialization of such functions in different units in more advanced economic systems; however, in the geographical and economic isolation in which much of Soviet agriculture is, in fact, carried on it might be impossible to employ mechanical and electrical equipment or up-to-date agronomical knowledge unless it is supplied within the farm framework.

To some extent the economic limitations of this situation are overcome by the growth of inter-collective farm enterprises of which there were 3,355 at the end of 1965.[54] Though this was a slight drop on the 3,431 in existence three years earlier,[55] this was on balance more than accounted for by the fall in the number of electric power stations which simply reflects the continuing switch to the national grid as a source of electricity and does not represent a decline in the available facilities. More than half of the total number of joint enterprises were building (and building material) units, but an increasing number—616 in 1965

—consists of joint egg and poultry factories which may have been too large to be managed within the orbit of individual collective farms.

Despite such substantial, and in principle promising, efforts to improve the efficiency of certain services by increasing their scale of operations, it may be true that Soviet agriculture is paying a heavy price for the general backwardness of the Soviet countryside, the insufficient public investment in the infrastructure of transport, communications and energy, etc. Much of this is probably inevitable, but there may also be a good deal of waste because of the political and technical ineptitude of the bureaucratic régime in its dealings with agricultural problems and with the peasants. Though these peculiarities of the régime will persist in the foreseeable future, their economic effects could diminish fairly quickly, if high overhead costs should be spread over a bigger volume of output.

Specialization and Concentration

The internal administration of the collective farms and, particularly since the conversion policy of the later 1950s and early 1960s, of the State farms is still bedevilled by the effects of collectivization on the technical efficiency of large-scale public farming.

The original collective farms of the 1930s generally continued many or most of the traits of the backward peasant economy on which they had been imposed by force, though the basic processes of grain cultivation and of that of a few other crops were gradually mechanized. Similarly, the much larger amalgamated farms of the 1950s and the State farms which absorbed so many collective farms after 1956, generally continued the old organization of production in a new guise.[56] After amalgamation with some neighbouring collective farms, each of the original collective farms often became a 'complex brigade', and even as late as 1962 over one-third of the total number of productive brigades in the collective farms—and well over one-half of all those in the Russian Federated Republic—fell into this category.[57] One of the aims of the Territorial Production Administrations introduced by Khrushchev in 1962 was to sort out the resulting muddle by enforcing a measure of specialization within the area of operation of the TPA, irrespective of whether the farms concerned were State farms or collective farms, and by stream-lining the operations of the farms themselves.[58]

The lumping together of mixed small-holdings into agglomerations of some sixty or seventy which were in turn combined by bureaucratic *fiat* into some hundreds, led to the multiplication of enterprises carried on frequently on a very small scale. Examples are quoted of collective farms engaged in fifteen to seventeen different enterprises each of which was carried out by a number of brigades. Thus on a collective farm in the Krasnodar area, almost all the thirty-three crops raised were spread evenly over the four brigades of which it was composed rather than being divided amongst them; the individual fields sown to some of the smaller crops were only six, four or even two hectares; livestock tends to be split up to such an extent that the individual herd may consist of only twenty or thirty cattle,[59] though the collective farm may figure in the statistics as the owner of hundreds or even thousands.

Despite the huge overall dimensions of most State and collective farms the size of their operative productive units is thus frequently too small for the use of modern machinery. At least one of the causes of this situation was the separation between the management of agricultural plant and machinery and general farm management, and the dissolution of the MTS was partly caused by the hope of doing away with it. However, this could not be done effectively at the level of the collective farm as a whole: if the only result of breaking up the MTS as an independent body was to maintain its factual independence as a 'tractor brigade' servicing the other brigades in the same way in which the MTS had serviced the collective farms in its area, the old weaknesses would be maintained and even compounded by the lower efficiency of the new system.

Integration could be made effective only at the level of the brigade, or even at that of its sub-division, the link. Hence the creation of 'mechanized' brigades (or links) in which any number of tractors, from two or three to eight or ten, with corresponding items of agricultural machinery, are combined with the manual labour force which previously followed up the work of

279

the MTS-operated machinery. In such a team the tractor drivers (of whom there are normally two for every tractor) would form the leading element; they are generally men, while the majority of the manual 'man-power' consists usually of women. By 1961 the system of mechanized brigades had been adopted on a large scale, particularly in the wide open regions of the south and east, and about 15 per cent of all collective farm brigades had been 'mechanized'.[60]

The twin methods prescribed for the rational organization of production within the individual public farm are specialization and concentration. *Specialization* may involve the separation of functions within the same enterprise, e.g. the milking of cows from the raising of herd replacements, while *concentration* combines into a single unit similar activities carried out by different persons and at different places within the same farm, e.g. the keeping of pigs or poultry in one brigade instead of two or three, or the planting of the total acreage of a minor crop planned for the whole farm in one field by one brigade rather than its distribution amongst two, three or even more.*

The Size of the Operative Unit
Although average figures for brigades of different types may not be very meaningful, they give a rough idea of the relative sizes of different types: in 1961 the simplest type, the non-mechanized field brigade, consisted on average of 14 adult workers and nearly 400 hectares of arable land, the mechanized field brigade had 121 workers and 1,120 hectares, the non-mechanized multi-purpose brigade 147 workers and 900 hectares and the mechanized multi-purpose brigade 183 workers and 1,700 hectares.[61]

In the literature on the question of the best size for the operative unit, the emphasis is mainly on the need for adequate minimum dimensions at the truly operative level, the cropping brigade or the livestock unit, in order to obtain the full economies of scale obtainable by the use of modern machinery. Thus the All-Union Agricultural Economics Research Institute

* In using the same slogan in a totally different setting, Dr S. Mansholt of the EEC Commission had to deny any intention to apply Soviet methods. (Address on *The Future Shape of Agricultural Policy*. Newsletter on the Common Agricultural Policy, Brussels, January 1968.)

recommended in 1963-64 that collective farm brigades should have minimum acreages of 400 to 600 hectares (and four to five tractors) for flax-growing farms in the non-black-earth belt of north-western European Russia, of 1,000 to 1,300 hectares (with six to seven tractors) for the grain-livestock farms of the non-black-earth belt, of 150 to 180 hectares (with five to six tractors) for specialized potato and vegetable farms, of 120 to 150 hectares for cotton growing farms, of 1,800 to 2,200 hectares for grain farms in the main grain-growing regions and of 2,200 to 2,600 hectares for grain and livestock farms in the south-east, western Siberia, Kazakhstan, etc. In favourable conditions, these figures may be increased by 50 to 100 per cent.[62]

For State farms, the following long-term optimum dimensions of grain growing farms are quoted. (The figures in brackets relate to each of their four or five divisions.) Northern Caucasus —steppe: 25,000-28,000 hectares per State farm (4,000-7,000 hectares per division); northern Caucasus—other districts: 16,000-22,000 (4,000-5,000) hectares; lower Volga—steppe: 24,000-36,000 (5,000-8,000) hectares; Urals and western Siberia —forest-steppe: 16,000-24,000 (4,000-6,000) hectares; Urals and western Siberia—steppe: 24,000-36,000 (5,000-8,000) hectares; Ukraine—southern steppe: 12,000-16,000 (4,000-5,000) hectares; Kazakhstan: 18,000-38,000 (5,000-8,000) hectares.[63]

The main difference between the recommendations for State farms and those for collective farms is the fact that for the former the *optimum* is put at a multiple of the technically appropriate *minimum*; thus in the Tseliny Krai the minimum acreage for a team with three tractors is given as 500 hectares, but the optimum size for a brigade as 3,000 to 3,500 hectares.[64] Similarly the economic minimum for a dairy herd with a herringbone parlour for sixteen cows is quoted as 150 cows (including 18 per cent not in milk) but the optimum as 400 to 1,000.[65]

How Much Too Large are Soviet Farms?

The preceding recommendations seem to have been made at the very moment when the policy of increasing the size of farms, particularly of State farms, was at its zenith and the opposite tendency was gaining strength. Their theoretical basis, at least for livestock farms, seems to have been an extremely simple and problematical production function which took a very sanguine view of the possibilities of expanding production within a given system of fixed assets and overestimated to an alarming extent the economies of scale available in the production of agricultural commodities; on this basis, it was assumed that greater volume of production would invariably lead to lower costs, though the fall in costs became increasingly less pronounced, but that rising transport costs would ultimately offset this benefit of expanding production.[66]

Administration was mentioned in this context not as a specific cost but only as a technical obstacle to unlimited expansion because of the 'difficulty' of administering giant farms. Thus the central leadership of the farm has the task of checking every month the performance of the productive units, the brigades and livestock farms. This involves no fewer than fifteen separate forms (thirteen in the collective farms) for brigades and mechanized links in field work;[67] needless to say, these forms have to be completed by the brigade management before they can be checked. Substantial as the cost of transport is in Soviet conditions, the steeply rising cost and declining efficiency of management and administration are likely to put narrower limits on optimum size in public farms than the cost of overcoming physical distance. René Dumont, whose ideological bias is by no means in favour of Western capitalism, observed 'that the administrative system of State farms grows more than in proportion to the scale of operations of the enterprise'.[68]

Even before Khrushchev's fall the policy of increasing the size of State farms, mainly through the conversion of collective farms, was put into reverse. Between 1962 and 1964 the average number of workers per State farm fell from 825 to 721, and two years later it had declined further to 650, which compares with 639 in 1958; the average sown area fell similarly from 10,100 hectares in 1962 to 8,600 in 1964 and 7,300 in 1966, compared with 8,700 in 1958 and the same trend was observable for average livestock holdings, though in this case the average numbers in 1966 remained higher than in 1958 at least for cows and other cattle.[69] Such movements are an eloquent comment on the preceding policy of hasty and excessive concentration of production in larger and larger units.

A detailed analysis of the State farms under the control of the Ministry of Agriculture at All-Union level in 1965 shows that 55 per cent of all such farms had, in fact, an arable acreage of less than 5,000 hectares, with 33.8 per cent between 5,000 and 20,000 and only 11.2 per cent over 20,000 hectares. The major part of the acreage on farms over 20,000 hectares was devoted to grain (obviously mainly in the 'new lands'), with sheep runs in second place. Almost all the cotton farms, almost two-thirds of the sugar-beet farms, 90 per cent of the fruit and vegetable orchards, over 80 per cent of the dairy farms, almost three-quarters of the beef farms and piggeries, 80 per cent of the poultry farms and 90 per cent of mixed other State farms were under 10,000 hectares.[70] Though these acreages are still very large, a determined effort to differentiate size according to enterprise and to review the results of the uncritical expansion policy of the previous decade is obviously in progress.

The pace at which the size of many thousands of quite complex farms was increased year by year between 1950 and 1962 would be incompatible with sound management, even if the reasoning behind the policy of farm amalgamation could be accepted in principle. It is, however, true that the size of the operative production unit, the brigade, increased much more slowly than that of the collective and State farms: between 1957 and 1961 the average arable acreage of the collective farms rose by 60.5 per cent and that of the average brigade by 21.7 per cent, though its able-bodied workers increased by 35.9 per cent.[71] Such increases remain, however, very considerable.

The first point to make in assessing the size of Soviet public farms is to distinguish the individual State and collective farm from the firm in a market economy as an economic unit. In the kaleidoscopic changes of functions between different organs of Soviet agriculture since 1953 the lines of demarcation between effective units of management have changed from time to time. There was undoubtedly a considerable degree of decentralization in the making of certain decisions from organs of central and regional government to public farms, and the increase in the size of these farms was a corollary of this process.

In a bureaucratic system of government, administrative functions tend to take precedence over economic functions, and the considerable administrative tasks with which State and collective farms are entrusted make it most unlikely that farm size in the Soviet Union will be determined simply by the criterion of reducing costs to a minimum or maximising the margin which public farms could earn over costs at existing producer price levels. To this extent, Soviet farms will remain too large, and the existence of a substantial nucleus of managing and technical personnel—amounting to six or seven persons for the average collective farm and perhaps three times that number for the average State farm—may well retain a limiting influence on the adaptation of farm size to optimum production costs.

On the other hand, there is some room for flexible arrangements in the allocation of functions between the public farm as a whole and its operative units which could well improve the size of the latter by adapting it more closely than in the past to the technological and economic optimum. The two critical factors will probably turn out to be labour and machinery, though the configuration, quality and quantity of land resources are bound to play an important part.

With the gradual abandonment of the original collective farm concept which involved the treatment of labour as the residuary legatee on collective farm income, the most important difference between State farm and collective farm has faded and the previously close connection between the operational results of the collective farm as a whole and the earnings of the individual collective farm member disappears. As this connection was outside the power of the individual to influence, this change creates the possibility of linking individual earnings much more closely to individual effort than in the past. This is stressed, reasonably enough, as the essential condition of efficient work.

The main operative unit, the brigade, appears still as an extremely unwieldy basis for this purpose. With a typical complement of more than a hundred workers it is only less unsuitable as a unit of labour management than the public farm as a whole. Such huge regiments of workers are, of course, the converse of the low productivity of labour in crop-raising and animal husbandry which reflects largely, though not wholly, the low level of mechanization in most branches of agriculture. While this remains true, the full use of the still scarce and dear resource of agricultural machinery is the main consideration in calculating the minimum size of the operative unit; but as the mechanical equipment available per worker increases and the efficient use of labour becomes increasingly important because it has to be paid a wage corresponding more closely to that of labour in other branches of the national economy, the unsuitability of such very large operating units for this purpose will become plain to see.

One method of reducing this unit to manageable dimensions without encroaching on the full utilization of agricultural machinery is the sub-division of the brigade into links which have been in turn praised and denounced. The link is defined as a team of workers within the brigade 'which carries out the main work for the cultivation of one or more crops on the land allotted to it and which is materially interested in obtaining a high crop at the lowest possible expenditure of labour and material inputs per unit of production'.[72] Whether in the fulness of time a wholly mechanized link will replace the brigade, or whether the number of brigades will be gradually increased and their size correspondingly reduced and brigades and links will co-exist with a different allocation of duties, cannot be profitably discussed at present. It is, however, possible to claim that in a modern agricultural set-up there will be no need, nor any room, for under-equipped and therefore over-staffed operating units of the size of the present brigades. These remnants of an era when agricultural labour was grossly under-paid and therefore wastefully employed will become obsolete as soon as the end of agricultural under-production and scarcity shows up their extravagant use of no longer plentiful labour resources.

It may not be possible to state in quantitative terms of acreage and livestock herds by how much the present size distribution of Soviet public farms exceeds the economic optimum. It is, however, possible to suggest some broad principles which might indicate the possible orders of magnitude. The size of State and collective farms as a whole may not be entirely dependent on purely economic considerations, because in certain circumstances they are not entirely limited to economic functions.

Nevertheless, they may gradually relinquish some of their auxiliary functions to specialized service organizations and their size should not exceed the point at which the economies of scale of administration are exhausted.

The operative unit may for some time be governed more by pressure for economies of scale in the utilization of scarce plant than by the benefits of better utilization of still abundant man-power. But agricultural man-power will ultimately become not only dearer than it has been in the past and also scarcer, as skilled labour capable of operating modern plant has always been in Soviet agriculture. As output per man rather than output per tractor or acre becomes the main consideration, the centre of gravity in Soviet agriculture will shift towards forms of organization in which the individual rural worker will no longer be a cog in a poorly fashioned wheel but will be enabled to make his maximum contribution as the most valuable factor of production.

The bureaucratic character of the whole system must make it problematical whether Dumont's optimistic vision of a federation of independent autonomous sections within a de-bureaucratized State farm[73] is attainable in Soviet Russia, but it is safe to assume that while there may be a significant reduction in the overall size of public farms and their operative units there is no need on economic grounds for a return to small scale farming even in the American sense of a fairly large territorial unit with a very small labour force. The sweeping thesis that the weaknesses of Soviet agriculture are essentially due to large-scale farming is no less idealogical and one-sided than the bureaucratic preference for size for its own sake. This applies, e.g., to Professor Th. Schultz's criticism of the 'absurd, bimodal structure of farm sizes, i.e. exceedingly large State and collective farms and tiny plot farms . . . based on big tractors and many hoes' and his suggestion of a radically different alternative: 'Suppose these plot farms were increased to no more than ten acres and suppose small hand (garden-type) tractors and complementary machines and equipment were made available: total agricultural production in the Soviet Union would rise sharply and chiefly of these farm products that are at present in short supply.'[74]

Ten acres was almost exactly the average sown area per peasant household in the NEP period and there were about twenty-five million of them. The mental experiment of supplying them with 25m. hand tractors and working machines not only ignores the political realities of the situation (which would not be unreasonable or unfair) but bypasses the central economic problem of under-development—the lack of the necessary material and human resources for modernization. The provision and servicing of the mechanical tools of modern agricultural production would have been quite beyond the ability of the non-agricultural sector, and the ability of the peasants to use them was almost totally lacking. There is a shortage of tractors in Soviet agriculture even now, and although the use of big—and probably excessively large—tractors reduces the number of operators needed to run them, there is a shortage of men capable of handling such equipment. To suggest, if only as a rhetorical flourish, that the heirs of generations of *muzhiks* could be transformed at short notice into a million-headed mass of miniature modern farmers capable of operating the resources which a highly industrialized society might put at their disposal, completely disregards the real problems with which the Soviet régime had to grapple.

On a different level, the relative shortage of feed grain remains the main brake on Soviet agriculture as a whole, and the transfer of some 150 to 200m. acres (sixty to eighty million hectares) to the new ten-acre small-holders would involve a shift in a high proportion of grain production to such farms. The existence of substantial economies of scale in grain production is one of the best established facts of modern agriculture. Despite absurdly high grain prices, the peasant agriculture of Western Europe, with average farm sizes of well above ten acres, is faced with a persistent and intractable structural problem of farm sizes. In Great Britain, commercial grain growing is heavily concentrated on farms exceeding, at the least, a minimum of 100 acres (40 hectares) and in Canada there is a strong trend towards holdings in excess of ten times that figure. However excessive the size of Soviet public farms, the movement towards larger farm sizes is almost universal in the agriculture of developed countries.

Perhaps the greatest obstacle to the successful application of

287

the experience of 'Western', and particularly of American, agriculture to Soviet conditions is the confusion of specific social and economic factors with purely technical practices which are, at least in principle, freely transferable between countries. This applies particularly to the relative values of the conventional factors of production, land, labour and capital.

For the American economist, the 'family farm' in the United States may appear as 'an excellent example of "small-holder" agriculture, with the typical farm having a two-man labour force'[75] supported by very large amounts of capital. To the extent that such a description is accurate, the underlying reality depends so thoroughly on the possibility of substituting capital for labour at their relative costs in the United States that it cannot be fairly applied to different conditions. The farming operation in the United States is 'supported' to such an extent by auxiliary industries that it is integrated into a chain of activities which may be dominated either by the suppliers of inputs (e.g. compounders in the case of livestock production) or by the processing and distributing interests (e.g. supermarkets) which prescribe the manner in which farming is conducted in considerable detail. The result may be a technically highly efficient system but one which is so completely dependent on the social and economic environment that it cannot even be adequately defined in terms of the farming operation—which forms a steadily diminishing portion of the whole—and still less be exported either in theory or in practice.

The Subsidiary Private Economy

The ingrained hostility of the Soviet régime towards the private plots of the peasants and rural workers probably owes as much to bureaucratic suspicion of the anarchic element of an imperfectly controlled market in an otherwise organized economy as to purely ideological factors. Perhaps the most outstanding features of the 'private sector' in agriculture are its vitality and pervasiveness: the number of persons 'interested' in it has been estimated by Soviet sources as one hundred million people or thirty million families, including perhaps as many as ten million non-agricultural workers and employees.[76] Probably the majority are not interested in producing for the market and a very high proportion of their output is consumed in kind. In 1962, the private sector was credited with 33.1 per cent of agricultural gross production but only with 16 per cent of market production; similar proportions applied to vegetable products (22.9 and 11 per cent) and livestock products (44.9 and 21 per cent).[77] Of the farm produce used directly for personal consumption, over 60 per cent of potatoes, 53 per cent of eggs, 43 per cent of melons, 35 per cent of vegetables and 30 per cent of the fruit was consumed in kind, most of it on private plots.[78]

The great economic importance of the private plots for their owners and for the economy as a whole, particularly as a barometer of the economic pressures in Soviet agriculture, is not diminished by the diminutive size of the individual holdings. On plots of one-quarter or one-half a hectare, with livestock numbered in ones or twos, the productivity of labour is inevitably low; however, at least until now, this inherent weakness of the private sector has been balanced partly by the higher intensity of labour and partly by the poor results of public agriculture.

A detailed, though necessarily speculative, calculation of the labour usage in the private sector by Nancy Nimitz estimates that in 1963 its share of total agricultural labour was as high as 42.1 per cent and its productivity only 71 per cent of labour in the public sector.[79] This calculation is, however, based on the relative shares of the public and private sectors in agricultural *gross* production; this is only permissible if the ratio between gross production and the value added (which is the true basis of comparisons involving the productivity of labour) is the same in both sectors. This is demonstrably not the case, for the share of material inputs in gross production is as high as 60 per cent in the State farms, 50 per cent in the collective farms and only 30 per cent in the private sector, mainly because of its low level of industrial supplies and depreciation charges.[80] The figures are not nearly detailed enough to justify precise calculations, but even if Miss Nimitz's high estimates of the labour used in the private sector are accepted, the value added per labour-hour on the private plots may be as high as, or even higher than, in the public sector.

In the past, the improvement in labour income from work in public agriculture has generally been achieved through higher prices for agricultural produce which have automatically at the same time increased the value of the private plots to the peasants, both as a source of income in kind and of money income. If the remuneration of labour in public agriculture is to rise without an increase in producer prices, the productivity of labour in the public sector must go up and the balance of supply and demand in the market of agricultural produce must improve, thus reducing the scarcity prices in the collective farm market. Until this happens, the tiny private plots of peasants and workers will retain their position in the economy of the rural population and in the supply of certain important foodstuffs (potatoes, vegetables, fruit, eggs and poultry) to the non-agricultural population. Though conditions in both respects are likely to change for the better during the next five or ten years, the change will probably be steady rather than dramatic, and even afterwards the importance of the private plots will fall gradually rather than disappear completely.[81]

There is nothing unusual in the existence of small potato patches and vegetable orchards with a few head of livestock or small farm animals to supplement the incomes which the rural proletariat can obtain from other sources. What makes the 'subsidiary private economy' of the peasants and workers in the Soviet Union so remarkable despite the minute scale on which it is operated is the still not completely forgotten link with the peasant economy of pre-collectivization days and the past failure of public agriculture to provide reasonable incomes to those employed in it and plentiful supplies of agricultural produce at reasonable prices to the population at large. In the last resort, neither failure was due to the size structure of Soviet farms as such, though its undoubted irrationality reflects the problems of Soviet agriculture as in a distorting mirror.

XI

The Balance of Soviet Agriculture

1. THE RETREAT FROM EXPLOITATION

The exploitation of the peasantry was the corner-stone of traditional Russian society; it preceded the development of an urban, industrial civilization and lasted as long as the Tsarist régime. Though the growing working class—which was itself only one step removed from the village—had its desperate grievances and good reason for revolutionary discontent, the peasant masses rather than the workers were the most under-privileged section of Russian society.

The oppression of the village population by Government, landlords, traders and money-lenders was not confined to Tsarist Russia; it was more the rule than the exception in the old world during the transition to modern industrialism. Agrarian discontent was, therefore, a normal feature of social and political life, and it depended on the historical context whether it formed part of the 'progressive' response to the advance of capitalism or whether it became, on the contrary, the ally of 'reactionary' forces which opposed this process for reasons of their own.

The relationship between the Soviet power and the peasants before the second agrarian revolution of 1929-32 was distinguished by the relative strength of the village after the overthrow of the Tsarist-protected landlord system. Despite its propagandist claims to represent the interests of workers *and* peasants, the Soviet power was at that time essentially confined to the towns; in the village it managed to extract a modest agricultural tax and to manipulate the prices at which the peasants could sell agricultural produce and had to buy industrial goods. The stubbornness with which the peasants defended the essence of the gains they had made in 1917 against these efforts doomed the market as the basis for their economic relations with the Government at the end of the NEP, and the bureaucratic character of the Soviet régime compelled the Communists to react by crude 'administrative' methods in the form of mass collectivization for the purpose of subjecting the peasants directly to the outside domination from which they had escaped during the Revolution.

Under Tsarism, the relations between the peasants and their oppressors fitted well enough into the classical Marxist scheme of classes and class conflicts. Russia's domestic history had turned for centuries on the endeavours of the landlords to hold on to the peasants as the ultimate source of their wealth and on the efforts of the peasants to shake off their yoke. With some degree of over-simplification it was possible to understand the Tsarist state as an auxiliary of the landlords in this struggle, though this did not make sufficient allowance for the independent position of the Tsarist bureaucracy *vis-à-vis* the main social classes.

The Soviet régime not only inherited the political dominance of Tsarism but after the NEP it came to occupy the strategic centre of post-revolutionary society. The social classes no longer acted directly on each other, but reacted in the first place to the initiative of the ruling bureaucracy, although the latter had to take some account of their interests in framing its policy. Perhaps the best illustration of this situation is furnished by the effects of Soviet agricultural policy in the Stalin era. Its worst excesses defy rational analysis and form part of the pathology of bureaucratic rule; from the point of view of Soviet society they represented pure loss. As far as they involved a genuine transfer of values, rather than their wanton destruction, the workers did not gain what the peasants lost, and the failure of the authorities to increase, or even to maintain, the volume of agricultural production depressed the working-class living standards at first even more than those of the peasants. If the village was bled white of resources, this was not done in order to provide the workers with cheap food: they had to pay the 'market price' for the necessaries of life, while the Government, as the guardian of the long-term interests of Soviet society, collected a swingeing turnover tax.

The crisis of under-investment, exhaustion and sheer unwillingness to operate a system which left the peasants as poor as before in a visibly progressing society clamouring for more agricultural produce was tackled after Stalin's death with great energy and considerable initial success. Khrushchev and his advisers may have under-estimated the need for sustained capital injections to make good the neglect of decades, and the extent to which producer prices lagged behind properly defined costs, but under the new leadership the régime went a long way towards restoring production incentives to the peasants and supplying agriculture with modern equipment of a basic type. What defeated Khrushchev in the end was his inability as a reformer to compel the régime to jump over its own shadow or to change the bureaucratic character of the system with its irrepressible tendency to suppress the symptoms of social and economic difficulties by administrative action instead of removing their causes.

As long as there was a crying need for more production of all agricultural commodities, while the prices of many products, particularly those of animal husbandry, were well below the cost of producing them, neither the direction of changes in producer prices nor their effect on the income and resource distribution between town and country presented serious problems: higher prices meant better production incentives, and rises in rural incomes reduced the patent exploitation of the village by the Government. The rising line of the producer price curve received its most recent upward twist in the measures decreed by Khrushchev's successors in the early part of 1965.

Since then Government procurement prices for agricultural produce have been sufficiently high to cover the heavily inflated costs of livestock products and to leave very substantial margins over costs in favourably situated regions or enterprises for their marketable surplus. The disincentive of low, and generally ridiculously inadequate, producer prices which operated strongly under Stalin, and at least for some products until the recent past, has thus been removed.

Does this mean that the agricultural population is no longer being exploited by the Soviet régime? The only satisfactory answer to such a question must be quantitative and this can only be done in the crudest fashion. One thing to bear in mind is the fact that the reorganization of Soviet agriculture during the last generation makes it necessary to modify the economic categories which were applicable to the relations between the peasants as commodity producers and the rest of society before 1929 which were, by and large, similar to those in capitalist market economies. (See Chaper IV, section 1.) The two most important changes have been that the link between producer prices and market prices has been severed completely for the major part of agricultural market output and that the connection between agricultural net income and investment in agriculture has become much looser.

The main instrument for separating the prices paid to producers in general, and to agricultural producers in particular, from market prices in public trade is the turnover tax. This tax still falls most heavily on the products of the food and light industries which are predominantly engaged in the processing of agricultural commodities. In the early 1960s products with an agricultural base amounted to about three quarters of the total 'consumption fund'.[1]

It can be estimated that the turnover tax imposed on the food and light industries amounted in 1965 to about 28,900m. rubles[2] out of a total turnover tax yield of 38,664m.[3] There is no easy way of allocating this sum to different products; a recent Western estimate of the net incidence of indirect taxes on wholesale prices assesses this at 10 per cent for food (after allowing for subsidies, such as State farm losses), 33 per cent for clothing and footwear and 45 per cent for consumer durables.[4] Turnover tax is no longer levied on some important food items, such as meat and dairy products, but this estimate may well be on the low side as far as food products are concerned. On balance it is likely that agriculture ultimately still bears a very substantial amount of turnover tax which continues to rest heavily on an agricultural base, but this amount is probably only a fraction of the enormous tribute to which it was subjected under Stalin.

However, there is a credit side to this account and it is formed by the investment in agriculture from public funds, excluding the investments of the collective farms themselves. This amounted in 1965 to 6,445m. rubles (including 5,295m. rubles productive investment) and in 1966 to 6,600m. (and 5,500m.) rubles.[5] 1966 was the first year of the new plan which provides for investments of 41,000m. rubles during 1966-70, an annual rate of over 8,000m. which must involve a figure of almost 10,000m. rubles for the final year of the plan. Though the total turnover tax levied on processed agricultural products may also not remain static during the years up to 1970, it is at least probable that the gap between the differential taxation imposed on the producers of agricultural produce through the high rate of turnover tax and the amount returned to agriculture through increased investment of public money in agriculture will narrow still further—provided the good intentions of the planners are put into practice despite the claims of other powerful interests on the available resources.

In view of the very crude assumptions on which these calculations have been based, no precise conclusion is possible, but there is a respectable case for the argument that in the not too distant future a position will be reached where the Government employs the weapon of differential indirect taxation of agricultural produce primarily for the purpose of financing a growing proportion of new investment in agriculture. As for direct taxation imposed on agriculture, this is probably rather below the differential rent in the sense of Ricardo-Marx, since it is levied only on the excess of net income over a certain basic percentage. It is thus no longer sheer sycophancy, as it would have been during the Stalin era, to claim that the exploitation of the 'village' by the 'towns' has, indeed, come to an end.

This broad conclusion can be confirmed by a comparison of wages and incomes. In 1966, the monthly money wage of State farm workers averaged 79.2 rubles[6] or about 80 per cent of the average wage in the economy as a whole and almost three-quarters of that in industry. Taking into account the income from private plots, the remuneration of agricultural workers was only marginally, if at all, lower than in most other comparable branches of economic activity. At the same time, the difference in earnings per *work day* between State farm workers and collective farmers had been gradually reduced, until by the mid-sixties it was eliminated in most parts of the Soviet Union excepting a large area of the Russian Federated Socialist

Republic.[7] The *annual* earnings of collective farmers from public argriculture remained well below those of State farm workers because of the lower number of days actually spent on collective farm work.

The current Five Year Plan for 1966-70 proposes an increase of 20 per cent in the average earnings of workers and employees and a rise of 35 to 40 per cent in the earnings of collective farm members from public agriculture. As the latter do not obtain much more than half their total income from the collective farms, it may be assumed that the total earnings of collective farm peasants are planned to rise somewhat, but not very much, faster than those of State farm workers, thus eliminating the remaining differences in earnings per man-day worked in some parts of the country. In view of the almost universal differential between the earnings of the agricultural population and wage rates in industry, the current—and still more the proposed—position in the Soviet Union is probably relatively better than in many other industrialized countries, though in absolute terms the earnings of the lower paid sections of the population remain low.

2. BEYOND THE CONQUEST OF SCARCITY

By abandoning the policy of massive exploitation of the peasantry, the Soviet authorities have removed the biggest single obstacle to a permanent solution of their agricultural difficulties. As they have at the same time stepped up their investment in land, buildings and agricultural machinery and scaled down their production targets to levels which appear very modest in view of current production trends, there is a good reason to believe that these targets may be reached, and for some commodities appreciably exceeded, in the course of the current Five Year Plan for 1966-70—provided, of course, that the investment plans are, in fact, carried out.

Such a moderately favourable development of agricultural supplies may be all that is needed to match the trend of internal demand. In recent years, the Soviet population spent on food-stuffs about sixty per cent of its total personal expenditure—in 1963, 68,123m. rubles out of a total of 110,297m. (including wear and tear).[8] These figures exclude expenditure on services and cannot, therefore, be compared with the position in advanced Western countries, but the proportion of personal expenditure on food in the Soviet Union remains remarkably high for an industrialized nation, and may therefore be expected to decline fairly steeply in the more or less near future.

It has been shown that for many agricultural commodities production trends in recent years, if continued, would be enough for supplies to overtake demand during the next decade. Although setbacks are no less possible in future than they have so often been in the past, the stage has been reached where the transition from chronic shortage to potential surplus is at least within the realm of the possible. When this time comes—and for many commodities it is not yet around the corner—the Soviet Union will have the agricultural production it needs, but at a heavy price. The resources employed in Soviet agriculture are very large, and the capital investment programme of the current Five Year Plan will make them still larger. In addition, with labour accounting for 40 per cent of total costs, the improvement in the incomes of peasants and State farm workers was directly reflected in cost levels which have risen particularly steeply since 1962. Whatever the qualifications needed in the interpretation of national average costs, in the Soviet Union even more than elsewhere, the extent of this process is at least illustrated by Table 66.

Between 1959 and 1965, the average cost of production in collective farms for the Soviet Union as a whole rose by over three-fifths for cotton, by almost one-half for milk and potatoes and by about one-third for grain, vegetables, live cattle and wool; only for sunflower seed was it on balance unchanged. In the State farms, the cost increase exceeded one-half for cereals, live cattle and wool; it was about two-fifths for cotton, three-tenths for milk and about one-sixth for sugar-beet, potatoes and vegetables.

As long as demand exceeds supply, it is clearly better to have sufficient, if dear, agricultural products, but it is at least a possibility that in a few years' time the Soviet Government may find themselves with adequate or even ample supplies of very expensive commodities on their hands. This applies particularly to

L

TABLE 66
Average costs of production of selected agricultural products, 1959–65
(in rubles per ton)

Product	Collective farms (actual costs)				State farms			
	1959	1962	1964	1965	1959	1962	1964	1965
Cereals*	36	37	44	48	40	46	50	66
Raw cotton	209	224	281	325	209	285	282	291
Sugar-beet	—	16	17	20	24	23	23	27
Sunflower seed	41	30	30	42	—	—	—	—
Potatoes	30	38	35	44	53	69	63	61
Vegetables	62	68	70	84	61	68	64	72
Cattle†	749	834	927	989	688	844	1,167	1,052
Pigs†	—	1,146	1,250	1,124	—	—	1,261	1,067
Sheep†	—	512	609	643	—	—	662	612
Milk	104	129	151	155	126	160	181	163
Wool	2,432	2,504	2,939	3,135	1,911	2,733	3,096	2,907

* Excluding maize.   † Increase in live weight.
Sources: All 1959 figures and State farm figures for 1962 from Lemeshev (ed.), *op. cit.* (in ch. VIII, 22), pp. 235ff. 1962: Collective farms: Narkhoz 1962, pp. 338f. 1964: Narkhoz 1964, pp. 396ff, 412f. 1965: Narkhoz 1965, pp. 408ff, 428f.

raw cotton, oil-seeds, sugar-beet and milk, for all of which output tends to increase somewhat faster than would be necessary in order to meet prospective home demand. There may be reasonable export prospects for cotton and sunflower seed (though not for sugar and milk products), but the time may not be too far off, when a ceiling on the production of some agricultural commodities may be considered necessary.

Though it may be regarded as somewhat premature to speculate on the prospects of solving problems which at the moment are barely more than tentative items on the agenda of the next-but-one Five Year Plan, this may help to elucidate the strength as well as the weakness of Soviet agriculture in the years to come.

With a few exceptions, Soviet agriculture is at present a system of vast resources subject to severe handicaps and producing a relatively low volume of output at generally high costs. Taking the official statistics at their face value, the output of vegetable products in the record year 1966 was 276 per cent of the 1913 level and that of livestock production 271 per cent, though this is a little difficult to reconcile with the rise in the volume of production of the main animal products. (Appendix,

Table I.) The sown area in 1966 was 206.8m. hectares[9] compared with 118.2m. in 1913, which suggests an overall increase in gross production per acre of 58 per cent in fifty-three years or about seven-eights per cent per year compound. During the half century 1916 to 1966, the number of cattle rose by 66 per cent, of pigs by 152 per cent and of sheep and goats by 46 per cent; the average increase in output per unit of productive livestock must have been, if anything, even lower than that in crop production per acre. It is impossible to compare the fixed capital employed in agriculture before the Revolution with current data, but between 1940 and 1966 it is claimed to have risen by 228 per cent including livestock and by 376 per cent excluding livestock.[10] As agricultural gross production during this period expanded according to the official calculation by 96 per cent, the capital: output ratio must have risen fairly sharply. Finally, the rural population (which is the best indication of the agricultural population for 1913) fell from 131.1m. in 1913[11] to 106.4m. in January 1967,[12] a decline of only 19 per cent in fifty-three years.

Making due allowance for the force of the historical hurricane which swept the Russian countryside twice, or even three times, during this period, perhaps the most moderate conclusion must be that the great revolution in agricultural yields and techniques which occurred in most other advanced countries in these years, and particularly since the Second World War, has bypassed the Soviet Union.

The deplorable effects of the microscopically small advances in most qualitative factors on the actual position of Soviet agriculture need no further elaboration. At the same time, a number of the conditions for catching up with Western Europe and America undoubtedly exist in the Soviet Union. There is no shortage of competent and trained scientific personnel, hence there is general awareness and specific knowledge of the basic techniques and their application to Soviet conditions; there is at last a real effort to supply capital goods and current industrial inputs in quantities commensurate to the task in hand; and finally there is a determination to pay the full price—perhaps even a very high price—to make the production of most or all agricultural commodities economically attractive at the high current cost levels. This means that there is enormous room for future improvement, and in many cases only a relatively narrow threshold to pass before this potential can be fully exploited in practice.

Once the incentive of scarcity for expansion in the face of steeply diminishing returns ceases to operate, the still proceeding vicious circle could be put into reverse with fairly dramatic effects. With seed accounting for perhaps as much as one-sixth of gross crop production, the withdrawal of marginal land from production compensated by higher crop yields could by itself help to increase the available net output; with maintenance absorbing a disproportionately high share of animal feed fed to inflated herds and flocks, a modest increase in yield per animal could provide the same output of livestock products from smaller herds at appreciably lower outlays in feed, thus reducing the need for high-cost marginal production of feed grains and fodder crops to the benefit of the overall efficiency of vegetable production. The systematic application of the well-known technological advances made in most branches of agriculture during the last generation could produce a chain-reaction of qualitative improvements in the practice of Soviet agriculture to the great benefit of agricultural producers and the Soviet economy as a whole.

Whether these highly desirable and technically perfectly possible developments will, in fact, take place depends mainly on the policy of the Soviet Government during the next five or ten years. In important respects, prospects are more favourable for a rational and therefore successful official policy towards the peasants and agriculture as a whole than at any time since the end of the NEP or, perhaps, since the October Revolution. However, this is far from identical with the certainty, or even the probability, that such a policy will be carried out consistently, or that a simple continuation of the measures at present in force will be sufficient for this purpose.

This applies, above all, to the price policy pursued by Khrushchev's successors since 1965. In a formal sense, the reintroduction of dual pricing for quotas and supplementary sales to public trade was probably a step backwards, open to the same objections as those raised against this practice in the later

1950s and best understood as a reaction against the defective and arbitary methods used in the administration of the single procurement prices since 1958. The drawbacks of the current practice would, of course, show up most clearly if and when supplies of some commodities began to exceed demand. Attempts to manipulate basic and supplementary delivery prices, which would be well in line with the method of operation of the Soviet bureaucracy, might well be no more successful in future than they have proved in the past.

Price policy is likely to become even more important than it is at present, if the current trend towards the decentralization of management in the Soviet economy continues; the arguments in its favour are, indeed, stronger in agriculture than in industry, for in this branch the economic penalty exacted for insufficient attention to special conditions and for tardy response to the needs of the day is much greater than in the more standardized world of large-scale industry. It is difficult to visualize that the Government would be willing to jeopardize its procurement programme of agricultural commodities while they are in short supply, but it might be much more willing to rely on purely economic pressures once the era of scarcity is over.

At the same time the large and growing involvement of the Government in the financing of agricultural investment makes it both improbable and inappropriate that the authorities would leave the free disposal of the valuable publicly-owned assets of large public farms to its managers any more than in industry. Nor does the distinction between State farms and collective farms appear particularly pertinent as a basis for deciding whether particular long-term investments should be financed by the individual enterprise—mainly through the procurement prices fixed by the Government—or from public funds. The reconciliation between national interests and those of the individual public farm may well become especially difficult within the framework of a policy of decentralized management. It is impossible to guess whether a solution to this intractable problem might be found on lines parallel to the classical landlord-tenant relationship in British agriculture, with the Government responsible for 'landlord's capital' in the British sense (i.e. land, buildings, etc.) as well as for the bulk of the increasingly costly agricultural plant and equipment, and the actual operators of the public farms responsible for their use. In this case it would presumably be necessary to impose a more flexible and accurate rent charge than the present agricultural tax which does take some account of the capital invested in the enterprise.

Perhaps the most significant shift in emphasis resulting from the conquest of scarcity of agricultural products would be the need for improvements in the system of agricultural marketing. Apart from the purely local collective farm market, this is still limited to the procurement system of public trade, with some subsidiary inter-collective farm organizations, State farm trusts, etc. Even at present, the rudimentary character of the marketing mechanism is one of the handicaps imposed on a rational agricultural system: 'The lack of specialization in high-yielding crops adapted to local areas is to an important extent related to the lack of improved systems of marketing and distribution and the lack of area specialization in livestock production. Little attention has been given to marketing and distribution problems and their effect on production specialization in the USSR.'[13] Though less true than when it was written, this statement is still substantially correct. However, quite apart from the close links between marketing and production, the marketing of agricultural produce will become an important task in its own right as soon as the permanent sellers' market in food comes to an end.

This catalogue of oustanding problems could well be substantially extended, though it is only fair to add that the list of unsolved issues in the agricultural policies of other countries is also impressive. If a survey of the present prospects of Soviet agriculture is bound to end on a note of uncertainty, dissatisfaction with past results need not induce pessimism about the future. The two congenital defects which Soviet Russia inherited from Russian history—a backward agriculture and a ruling bureaucracy—have both proved desperately resistant to the healing effects of time, but there is good reason to believe that the backwardness of agriculture will be shaken off sooner than the bureaucratic character of the Soviet régime.

Appendix

TABLE I
*Index of agricultural gross
production* (1913=100)

Year	Total	Crops	Livestock products	Year	Total	Crops	Livestock products
1913	100	100	100	1945	86	93	72
1917	88	81	100	1946	95	100	87
1920	67	64	72	1947	122	140	89
1921	60	55	67	1948	136	158	96
				1949	140	156	109
1922	75	75	73	1950	140	151	118
1923	86	84	88	1951	130	133	126
1924	90	82	104	1952	142	148	129
1925	112	107	121				
1926	118	114	127	1953	146	148	141
1927	121	113	134	1954	153	153	153
1928	124	117	137	1955	170	175	160
				1956	193	201	177
1929	121	116	129	1957	197	198	196
1930	117	126	100	1958	218	227	205
1931	114	126	93	1959	219	215	221
1932	107	125	75	1960	224	226	219
1933	101	121	65	1961	230	230	229
1934	106	125	72	1962	233	229	235
1935	119	138	86	1963	216	209	221
1936	109	118	96	1964	247	270	217
1937	134	150	109				
1938	120	120	120	1965	252	247	254
1939	121	125	119	1966	276	276	271
1940	141	155	114	1967	279	—	—
				1968	—	—	—
1941	87	—	—				
1942	54	—	—				
1943	52	—	—				
1944	76	—	—				

Note: The index figures for 1917-39 are based on gross production in 1913 within the borders of the Soviet Union on 17 September, 1939; for later years, the base is gross production in 1913 within contemporary borders. Sources: 1917-40 and 1945-58: Selkhoz, p. 79. 1941-44: Caluculated from *Istoriya velikoi otechestvennoi voiny* . . . 1941-45 (1965), VI, pp. 63ff. 1959-62: Narkhoz 1962, p. 226. 1963-66: SSSR v tsifrakh (1966), p. 80. 1967: provisional; calculated from *Pravda*, 25 January, 1968.

TABLE II

Gross production of major crops (m. metric tons)

Year	Grain	Raw cotton	Sugar-beet	Sunflower seed	Flax	Potatoes	Vege-tables
1913 A	86.0	0.74	11.3	0.75	0.40	31.9	5.5
B	76.5	0.74	10.9	0.74	0.33	23.3	—
1923	56.9	0.14	2.6	—	0.22	—	—
1924	51.8	0.36	3.4	—	0.30	—	—
1925	72.5	0.54	9.1	2.22	0.30	38.6	—
1926	76.8	0.54	6.4	1.54	0.27	43.0	—
1927	72.3	0.72	10.4	2.13	0.24	41.2	—
1928	73.3	0.82	10.1	2.13	0.32	46.4	—
1929	71.7	0.86	6.3	1.76	0.36	45.6	—
1930	83.5	1.11	14.0	1.63	0.44	49.4	—
1931	69.5	1.29	12.0	2.51	0.55	44.8	—
1932	69.9	1.27	6.6	2.27	0.50	43.1	—
1933	89.8	1.32	9.0	—	0.56	—	—
1937	120.3	2.58	21.9	—	0.57	65.6	—
1940	95.5	2.24	18.0	2.64	0.35	75.9	13.7
1941	56.3	2.49	2.0	0.90	—	26.6	—
1942	29.6	1.32	2.2	0.29	—	23.5	—
1943	29.6	0.72	1.3	0.79	—	35.9	—
1944	48.7	1.14	4.1	1.03	—	54.6	—
1945	47.3	1.16	5.5	0.84	0.15	58.3	10.3
1950	81.2	3.54	20.8	1.80	0.26	88.6	9.3
1953	82.5	3.85	23.2	2.63	0.16	72.6	11.4
1954	85.6	4.20	19.8	1.91	0.22	75.0	11.9
1955	103.7	3.88	31.0	3.80	0.38	71.8	14.1
1956	125.0	4.33	32.5	3.95	0.52	96.0	14.3
1957	102.6	4.21	39.7	2.80	0.44	87.8	14.8
1958	134.7	4.37	54.4	4.63	0.44	86.5	14.9
1959	119.5	4.64	43.9	3.02	0.36	86.6	14.8
1960	125.5	4.29	57.7	3.97	0.43	84.4	16.6
1961	130.8	4.52	50.9	4.75	0.40	84.3	16.2
1962	140.2	4.30	47.4	4.80	0.43	69.7	16.0
1963	107.5	5.21	44.1	4.28	0.38	71.8	15.2
1964	152.1	5.28	81.2	6.06	0.35	93.6	19.5
1965	121.1	5.66	72.3	5.46	0.48	88.7	17.6
1966	171.2	5.98	74.0	6.15	—	87.9	17.9
1967	147.9	6.00	87.1	6.61	—	95.5	20.5
1968	169.2	6.00	93.6	6.64	—	101.6	18.5
1909–13 A	72.5	0.68	10.0	—	0.32	30.6	—
B	65.2	0.68	9.7	—	0.26	22.4	—
1928–32	73.6	1.04	9.8	1.83	0.43	45.9	—
1933–37	72.9	1.84	14.6	1.27	0.32	49.8	—

Year	Grain	Raw cotton	Sugar-beet	Sunflower seed	Flax	Potatoes	Vege-tables
1938–40	77.9	2.51	15.8	2.03	0.32	47.9	—
1949–53	80.9	3.49	21.1	2.05	0.23	75.7	10.0
1954–58	110.3	4.20	35.5	3.42	0.40	83.4	14.0
1959–63	124.7	4.60	48.8	4.16	0.40	79.4	15.8
1961–65	130.3	5.0	59.2	5.07	0.43	81.6	16.9

Note: A in respect of 1913 and 1909-13 refers to production within the contemporary borders, B to production within the borders of the Soviet Union in September 1939.

Sources: 1923-37: P. I. Liashchenko, *Istoria Narodnovo Khoziaistva* SSSR (1956) III, pp. 134f, 257, 371, with grain yields for 1933 and 1937 on the basis of 'biological yield'; supplemented for sunflower seed and potatoes by Jasny, *Socialized Agriculture*, etc. Other years up to 1958: Selkhoz, *passim* except for 1941-44 which are calculated from index figures in *Istoriya velikoi otechestvennoi voiny*, *l.c.* (in App. Table I). 1959-62: Narkhoz 1962, pp. 234f. 1962-4 Narkhoz 1964, p. 249. 1965-6 Narkhoz 1965, p. 262 and SSSR *v tsifrakh 1966*, pp. 82ff. 1967 and 1968: *Pravda*, January 26, 1969.

TABLE III
Gross production of main livestock products

Year	Meat (and fat) deadweight, m. tons	Milk m. tons	Eggs '000 m.	Wool '000 t.
1913 A	5.0	29.4	11.9	192
B	4.1	24.8	10.2	180
1928	4.9	31.0	10.8	182
1929	5.8	29.8	10.1	183
1930	4.3	27.0	8.0	141
1931	3.9	23.4	6.7	98
1932	2.8	20.6	4.4	69
1933	2.3	19.2	3.5	64
1934	2.0	20.8	4.2	65
1935	2.3	21.4	5.8	79
1936	3.7	23.5	7.4	99
1937	3.0	26.1	8.2	106
1938	4.5	29.0	10.5	137
1939	5.1	27.2	11.5	150
1940 B	3.9	26.6	10.2	151
1940 A	4.7	33.6	12.2	161
1941	4.1	25.5	9.3	161
1942	1.8	15.8	4.5	126
1943	1.8	16.4	3.4	105
1944	2.0	22.2	3.5	109
1945	2.6	26.4	4.9	111
1946	3.1	27.7	5.2	119
1947	2.5	30.2	4.9	125

Year	Meat (and fat) deadweight, m. tons	Milk m. tons	Eggs '000 m.	Wool '000 t.
1948	3.1	33.4	6.6	146
1949	3.8	34.9	9.1	163
1950	4.9	35.3	11.7	180
1951	4.7	36.2	13.3	192
1952	5.2	35.7	14.4	219
1953	5.8	36.5	16.1	235
1954	6.3	38.2	17.2	230
1955	6.3	43.0	18.5	256
1956	6.6	49.1	19.5	261
1957	7.4	54.7	22.3	289
1958	7.7	58.7	23.0	322
1959	8.9	61.7	25.6	357
1960	8.7	61.7	27.4	357
1961	8.7	62.6	29.3	366
1962	9.5	63.9	30.1	371
1963	10.2	61.2	28.5	373
1964	8.3	63.3	26.7	341
1965	10.0	72.6	29.1	357
1966	10.7	76.0	31.7	371
1967	11.5	79.9	33.9	395
1968	11.6	82.1	35.5	413
1909–13 A	4.8	28.8	11.2	192
B	3.9	24.1	9.5	180
1928–32	4.3	26.3	8.0	135
1933–37	2.7	22.2	5.8	83
1938–40	4.5	27.6	10.8	146
1949–53	4.9	35.7	12.9	198
1954–58	6.9	48.7	20.1	272
1959–63	9.2	61.9	28.2	365
1961–65	9.3	64.7	28.7	362

Note: For the meaning of A and B see Table II.
Sources: Up to 1958: Selkhoz, pp. 328f except for 1941-44, which are calculated from index figures in *Istoriya velikoi otechestvennoi voiny, l.c.* (in App., Table I). 1959-62: Narkhoz 1962, p. 309. 1963-64: Narkhoz 1964, p. 361. 1965-68: SSSR *v tsifrakh 1966* and *Pravda*, January 26, 1969.

TABLE IV

Livestock numbers (m. head)

Year (Jan. 1)	Cattle (total)	Cows	Pigs	Sheep	Goats	Horses	Poultry
1916 A	58.4	28.8	23.0	89.7	6.6	38.2	—
B	51.7	24.9	17.3	82.5	6.2	34.2	—
1922	35.0	—	8.6	52.5	1.2	20.2	—
1924	57.7	—	19.3	98.4		25.0	—
1926	64.0	—	20.3	121.7		28.8	—
1928 A	66.8	33.2	27.7	104.2	10.4	36.1	—
B	60.1	29.3	22.0	97.3	9.7	32.1	—
1929	58.2	29.2	19.4	97.4	9.7	32.6	—
1930	50.6	28.5	14.2	85.5	7.8	31.0	—
1931	42.5	24.5	11.7	62.5	5.6	27.0	—
1932	38.3	22.3	10.9	43.8	3.8	21.7	—
1933	33.5	19.4	9.9	34.0	3.3	17.3	—
1934	33.5	19.0	11.5	32.9	3.6	15.4	—
1935	38.9	19.0	17.1	36.4	4.4	14.9	—
1936	46.0	20.0	25.9	43.8	6.1	15.5	—
1937	47.5	20.9	20.0	46.6	7.2	15.9	—
1938	50.9	22.7	25.7	57.3	9.3	16.2	—
1939	53.5	24.0	25.2	69.9	11.0	17.2	—
1940	47.8	22.8	22.5	66.6	10.1	17.7	—
1941	54.5	27.8	27.5	79.9	11.7	21.0	255.7
1942	31.6	15.0	8.3	70.5		10.0	—
1943	28.3	13.9	6.1	61.4		8.2	—
1944	33.8	16.5	5.5	63.2		7.8	—
1945	44.1	21.4	8.8	70.5		9.9	—
1946	47.6	22.9	10.6	58.5	11.5	10.7	—
1947	47.0	23.0	8.7	57.7	11.6	10.9	—
1948	50.1	23.8	9.7	63.3	13.5	11.0	—
1949	54.8	24.2	15.2	70.4	15.2	11.8	—
1950	58.1	24.6	22.2	77.6	16.0	12.7	—
1951	57.1	24.3	24.4	82.6	16.4	13.8	292.8
1952	58.8	24.9	27.1	90.5	17.1	14.7	—
1953	56.6	24.3	28.5	94.3	15.6	15.3	—
1954	55.8	25.2	33.3	99.8	15.7	15.3	400.4
1955	56.7	26.4	30.9	99.0	14.0	14.2	—
1956	58.8	27.7	34.0	103.3	12.9	13.0	432.1
1957	61.4	29.0	40.8	108.2	11.6	12.4	449.7
1958	66.8	31.4	44.3	120.2	9.9	11.9	—
1959	70.8	33.3	48.7	129.9	9.3	11.5	482.8
1960	74.2	33.9	53.4	136.1	7.9	11.0	514.3
1961	75.8	34.8	58.7	133.0	7.3	9.9	515.6
1962	82.1	36.3	66.7	137.5	7.0	9.4	542.6
1963	87.0	38.0	70.0	139.7	6.7	9.1	550.4

Year (Jan. 1)	Cattle (total)	Cows	Pigs	Sheep	Goats	Horses	Poultry
1964	85.4	38.3	40.9	133.9	5.6	8.5	448.9
1965	87.2	38.8	52.8	125.2	5.5	7.9	456.0
1966	93.4	40.1	59.6	135.3		—	—
1967	97.1	41.2	58.0	141.0		—	—
1968	97.2	41.6	50.9	144.0		—	—
1969	95.7	41.2	49.0	146.1		—	—

Note: For the meaning of A and B see Table II.

Sources: 1919-61: Selkhoz, p. 263, except for the (not fully comparable) N.E.P. years and for 1942-45 which are calculated from index figures in *Istoriya velikoi otechestvennoi voiny, l.c.* (in App. Table I). 1961-65: Narkhoz 1964, pp. 352, 360 and Narkhoz 1965, p. 307. 1966-69: *Pravda*, January 26, 1969.

References

The most frequently quoted standard statistical compilations are: *Selskoie Khoziastvo* SSSR (statisticheski sbornik), Moscow 1960, quoted as *Selkhoz,* and *Narodnoie Khoziastvo* SSSR *v . . . godu,* the statistical year book, quoted as *Narkhoz* and the relevant year.

CHAPTER II
1. (Sir) W. S. Churchill, *War Memoirs* IV, 447f (1951).
2. Paul W. Blackstock and Bert F. Hoselitz (editors): *The Russian Menace to Europe by Karl Marx and Friedrich Engels* (1952), pp. 218ff, 275ff.
3. *Capital,* vol. I, ch. 24, section 6.

CHAPTER III
1. The author has provided at least an outline analysis for the Tsarist bureaucracy in *The Ruling Servants* (1961), chapter 8, and for the period 1917-39 in *Soviet Russia: Anatomy of a Social History* (1941). The condensed precis in the text summarizes the main results of this analysis without attempting to reproduce the evidence on which it is based.
2. *The Ruling Servants,* pp. 85ff.
3. *Ibid.,* pp. 163ff.
4. *Sochinenie* (4th ed.), XXXVI, 553
5. Quoted by M. Miller, *The Rise of the Russian Consumer* (1965), p. 90 from *Pravda,* April 24, 1964.
6. M. Lewin, *Russian Peasants and Soviet Power* (English translation, 1968), pp. 119ff.
7. Strauss, *Soviet Russia* (1941), p. 115.
8. *Piatiletni Plan* (1929), I, 90.
9. *Istoriya Kolkhoznovo Prava* (1958), 2 vols.
10. *Selkhoz,* pp. 57f, *Narkhoz 1964,* p. 391.
11. N. S. Khrushchev, *Stroitelstvo kommunizma v* SSSR *i razvitie selskovo khoziaistva,* 8 volumes (1962-64), I, 63ff.
12. *Sovremenny etap kommunisticheskovo stroitelstva i zadachi partii,* etc. (1962) pp. 368f where Khrushchev makes this point in his report to the Central Committee of the Communist Party in March 1962.
13. *Ibid.,* p. 370.
14. Howard R. Swearer, *Agricultural Administration under*

Khrushchev in R. D. Laird (editor), *Soviet Agricultural and Peasant Affairs* (1964), p. 29.

15. Swearer, *ibid.*, p. 16.

16. Lewin, *op. cit.* (in III, 6) notes this feature as a characteristic *modus operandi* of Soviet State and Communist Party as early as the 1920s. See also A. Nove, *Soviet Agriculture Marks Time,* reprinted in *Was Stalin really Necessary?* (1964), pp. 155ff. Khrushchev criticizes under the same name a different but related attitude. On a more technical level, the frequent neglect of the 'idea of complementarity of inputs' noted by A. Kahan, *Changes in Agricultural Productivity in the Soviet Union.* Report of the Conference on International Trade and Canadian Agriculture, Banff (1966), pp. 361f is a close parallel from the field of agricultural planning.

17. *Op. cit.* (in III, 12) p. 17; compare his frequent detailed criticisms of the system, e.g. in *op. cit.* (in III, 11), VI, 171ff.

18. A. Kahan, *Soviet Statistics of Agricultural Output* in Laird, *op. cit.* (in III, 14) p. 136.

19. For this disreputable episode see N. Jasny, *Socialized Agriculture in the* USSR (1949), pp. 725ff, and for the more recent shift from so-called barn yield to combine yield see P. Dibb, *The Economics of the Soviet Wheat Industry* (1966), p. 72 and *Narkhoz 1964,* p. 814.

20. Khrushchev, *op. cit.* (in III, 11), I, 38.

21. *Ibid.*

CHAPTER IV

1. *Sovietskoie Narodnoie Khoziaistvo 1921-1925* (1960), p. 223.

2. W. W. Rostow, *The Stages of Economic Growth* (1960), p. 38.

3. For the important distinction between 'gross' and 'net' marketing see J. F. Karcz, *Thoughts on the Grain Problem* in *Soviet Studies* 18 (1967), pp. 399ff.

4. K. P. Obolenski, (ed.) *Voprosy ratsionalnoi organizatsii i ekonomiki selskokhoziaistvennovo proizvodstva* (1964), pp. 188ff, 331. For a more sophisticated analysis see A. G. Koriagin, *Vosproizvodstvo v sotsialisticheskom selskom khoziaistve* (1966), ch. III.

5. *Piatiletni Plan* (see III, 8) I, 105, 133f.

6. *Ibid.*, pp. 132f, 139.

7. For a critical discussion of the Soviet adaptation of the Marxian scheme see P. J. D. Wiles, *The Political Economy of Communism* (1962), pp. 280ff.

8. *Direktivy* KPSU *i Sovietskovo gosudarstva po khoziaistvennym voprosam* I, 39.
9. FAO, *Monthly Bulletin of Agricultural Economics and Statistics,* February 1964.
10. J. W. Mellor, *Increasing Agricultural Production in the Early Stages of Economic Development* in E. O. Haroldson (ed.), *Food: One Tool in International Economic Development* (1962), p. 227.
11. *Ibid.,* p. 224.
12. *Piatiletni Plan* I, 141. Jasny, *op. cit.* (in III, 19) p. 791.
13. *Op. cit.* (in II, 2) p. 224.
14. See note III, 8.
15. *Op. cit.* (in IV, 2), p. 24.
16. *Narkhoz 1962,* pp. 7f.
17. *Selkhoz,* p. 450, *Narkhoz 1964,* p. 419.
18. *Narkhoz 1964,* pp. 543f.
19. *Selkhoz,* pp. 460f.
20. Koriagin, *op. cit.* (in IV, 4), p. 221.
21. *Piatiletni Plan,* I, 106.
22. A. N. Yefimov, *Ekonomika* SSSR *v poslevoienny period* (1962), p. 427.
23. *Op. cit.* (in IV, 1), p. 55.
24. *Narkhoz 1964,* pp. 580ff.
25. *Ibid.,* p. 153.
26. *Ibid.,* p. 661, *Selkhoz,* pp. 86, 202.
27. *Narkhoz 1962,* p. 547.
28. *Ibid.,* p. 545.
29. Dibb, *op. cit.* (in III, 19), pp. 40ff.
30. *Narkhoz 1964,* p. 661.
31. Dibb, *op. cit.* (in III, 19), p. 46.
32. *Ibid.,* p. 46.

CHAPTER V
1. *Works* (English edition) xxi, 1, p. 133.
2. Quoted by Louis Fischer, *The Life of Lenin* (1965), p. 151.
3. *Sochinenie,* vol. xxix, p. 331.
4. *Sov. Nar. Khoz.* (see IV, 1), pp. 234f.
5. *Ibid.,* p. 236.
6. N. A. Voznessenski, *The Economy of the* USSR *during World War Two* (Washington 1948), p. 54.
7. Strauss, *op. cit.* (in III, 1 and 7), p. 65.
8. CPSSSR *v rezolutsiakh,* vol. I, pp. 563f.

9. *Sov. Nar. Khoz.* (see IV, 1), p. 239.
10. *Ibid.*, p. 245.
11. *Narkhoz 1962*, p. 7.
12. *Ibid.*, p. 117.
13. *Itogi desiatiletia sovietskoi vlasti v tsifrakh 1917-1927*, p. 223.
14. *Sov. Nar. Khoz.* (see IV, 1), p. 413.
15. *Ibid.*, p. 416.
16. *Ibid.*, p. 447.
17. *Ibid.*, pp. 233, 251.
18. *Ibid.*, p. 316.
19. *Selkhoz*, p. 128.
20. On the basis of the same statistics, Lyashchenko divided the peasants into five groups; proletarians (no means of production), semi-proletarians (up to 200 rubles), small commodity producers (200-800 rubles), prosperous peasants (800-1,600 rubles), and small capitalists (over 1,600 rubles means of production). *Istoriya Narodnovo Khoziaistva* SSSR, vol. iii, pp. 238ff.
21. *Sov. Nar. Khoz.* (see IV, 1), p. 230.
22. *Ibid.*, p. 270, Lewin, *op. cit.* (in III, 6), pp. 41ff.
23. Stalin, *Problems of Leninism* (1940 edition), pp. 205f. For a discussion of the source of Stalin's figures see Karcz, *op. cit.* (in IV, 3).
24. *Op. cit.* (in IV, 1), pp. 330ff.
25. *Ibid.*, p. 270.
26. *Kollektivizatsia 1927-1935* (1957), p. 20.
27. Stalin, *op. cit.* (in V, 23), II, 129. Lewin, *op. cit.* (in III, 6), chapter 9.
28. N. Spulber, *Soviet Strategy for Economic Growth* (1964), pp. 79, 144 and the same (editor), *Foundations of Soviet Strategy for Economic Growth* (1964), pp. 283ff.
29. Wiles, *op. cit.* (in IV, 7), p. 32.
30. For detailed discussions of the whole controversy see E. H. Carr, *Socialism in One Country 1924-1926*, vol. I (1958), pp. 189ff, A. Ehrlich, *The Soviet Industrialization Debate 1924-1928* (1960), Spulber, *op. cit.* (in V, 28) and Lewin, *op. cit.* (in III, 6), pp. 132ff.
31. *Economic Equilibrium in the System of the* USSR (1927) in Spulber (ed.), *op. cit.* (in V, 28), p. 135.
32. *Ibid.*, p. 142.
33. *Ibid.*, p. 171.
34. *Op. cit.* (in IV, 1), p. 414.
35. *The New Economics* (English translation, 1965), p. 40.

CHAPTER VI

1. K. Kautsky, *Der Bolschewismus in der Sackgasse* (1930).
2. Strauss, *op. cit.* (in III, 7), pp. 156f.
3. M. Dobb, *Soviet Economic Development since 1917* (1960 ed.), pp. 206f.
4. The *locus classicus* for the drift in Soviet policy during 1926-27 is Lewin, *op. cit.* (in III, 6), chapter 7. See also the interesting discussion in *Soviet Studies,* vols. 16-8.
5. *Op. cit.* (in IV, 1), pp. 282ff.
6. *Selkhoz,* p. 41.
7. *Piatiletni Plan,* I, 90.
8. *Op. cit.* (in IV, 1), p. 305.
9. *Selkhoz,* pp. 56f.
10. Lewin, *op. cit.* (in III, 6), p. 443.
11. *Op. cit.* (in V, 26), p. 18.
12. Lewin, *op. cit.* (in III, 6), p. 514.
13. *Ibid.,* p. 499.
14. Lyashchenko, *op. cit.* (in V, 20), p. 368.
15. *Selkhoz,* p. 56.
16. *Ibid.,* p. 130.
17. Y. A. Moshkov, *Zernovaya problema v gody sploshnoi kollektivizatsii,* etc. (1966), p. 230. This supersedes Jasny's estimates in *op. cit.* (in III, 19), p. 751.
18. Moshkov, *op. cit.* (in VI, 17), p. 225 for 1932 and pp. 132f. for a detailed breakdown of the utilization of procurements in 1931-32.
19. Moshkov, *ibid.,* pp. 216ff gives some particulars of the methods used by the authorities. See also D. G. Dalrymple, *The Soviet Famine of 1932-34* in *Soviet Studies,* 15 (1964), pp. 250ff.
20. Strauss, *op. cit.* (in III, 7), pp. 237ff.
21. *Selkhoz,* p. 42.
22. *Ibid.,* p. 49.
23. *Ibid.,* p. 130.
24. *Ibid.,* pp. 48, 60.
25. *Ibid.,* pp. 56ff.
26. *Ibid.,* p. 58.
27. *Ibid.,* p. 53.
28. *Ibid.,* p. 74.
29. *Aid to the Kolkhoz Village* (1939) published by the People's Commissariat for Agriculture and quoted by Bienstock, Schwarz and Yugow, *Management in Russian Industry and Agriculture* (1944), pp. 153f.

30. *Op. cit.* (in III, 12), p. 373.
31. Bienstock, Schwarz and Yugow, *op. cit.* (in VI, 29), p. 156.
32. *Selkhoz,* pp. 56, 387.
33. *Ibid.,* p. 387.
34. Jasny, *op. cit.* (in III, 19) p. 403, and for the earlier position, Moshkov, *op. cit.* (in VI, 17), pp. 182ff.
35. Bienstock, Schwarz and Yugow, *op. cit.* (in VI, 29), pp. 165ff.
36. *Golovokhruzhenie ot uspiekhov* (1930), p. 6; see also *op. cit.* (in V, 26), pp. 409f.
37. *Selkhoz,* p. 128.
38. For a detailed summary of the Model Statute see Lyashchenko, *op. cit.* (in V, 20), III, pp. 411ff.
39. Jasny, *op. cit.* (in III, 19) p. 341.
40. *Ibid.,* pp. 356f.
41. Quoted by Jasny, *ibid.,* p. 32.
42. *Selkhoz,* p. 64.
43. J. G. Chapman, *Real Wages in Soviet Russia since 1928* (1963), p. 105.
44. *Pravda,* March 11, 1939.
45. Moshkov, *op. cit.* (in VI, 17), p. 221.
46. N. Jasny, *Soviet Industrialization 1928-1952* (1961), p. 173.— In his *Essays on the Soviet Economy* (1962), p. 101, Jasny estimated the average income of the agricultural population (in 1926-27 rubles per head) at 112.8 in 1927-28 and 71.1 in 1938.
47. N. Jasny, *Soviet Industrialization,* etc., p. 227, and *The Soviet 1956 Statistical Handbook: A Commentary* (1957), p. 41.
48. A. Bergson and H. Haymann Jnr., *Soviet National Income and Product, 1940-1948* (1954), p. 59.
49. *Pravda,* March 11, 1939.
50. *Piatiletni Plan* II, p. 270.
51. Obolenski, *op. cit.* (in IV, 4), pp. 8f.

CHAPTER VII
1. *Narkhoz 1962,* p. 8 (footnote) quotes Khrushchev's letter published in *Mezhdunarodnaia Zhizn,* 1961, No. 12, p. 8. See also J. A. Newth, *The Soviet Population: Wartime Losses and the Post-war Recovery* in *Soviet Studies,* vol. 15, pp. 345ff.
2. *Narkhoz 1962,* p. 8.
3. Voznesensky, *op. cit.* (in V, 6), p. 55.
4. Fedor Belov, *The History of a Soviet Collective Farm* (1956 ed.), pp. 107f.
5. Voznesensky, *op. cit.* (in V, 6), pp. 24, 94.

6. D. I. Shigalin, *Narodnoie Khoziaistvo v Period Velikoi Otechestvennoi Voiny* (1960), p. 197.
7. Voznesensky, *op. cit.* (in V, 6), p. 20.
8. Shigalin, *op. cit.* (in VII, 6), p. 204, Chadayev, *Ekonomika SSSR v Period Velikoi Otechestvennoi Voiny* (1965), p. 358.
9. Voznesensky, *op. cit.* (in V, 6), p. 96.
10. Chadayev, *op. cit.* (in VII, 8), p. 360.
11. *Ibid.*, p. 329.
12. Benediktov, in *Sots. Selskoie Khoziaistvo* (January 1, 1946) quoted by Jasny, *Soviet Industrialization*, etc. (VI, 46), p. 288.
13. Belov, *op. cit.* (VII, 4), pp. 118, 121.
14. Voznesensky, *op. cit.* (in V, 6), p. 20.
15. *Selkhoz*, pp. 88ff.
16. Voznesensky, *op. cit.* (in V, 6), p. 83.
17. Chapman, *op. cit.* (in VI, 43), p. 23.
18. *Ibid.*, p. 14.
19. Voznesensky, *op. cit.* (in V, 6), p. 75.
20. Shigalin, *op. cit.* (in VII, 6), pp. 178f.
21. For an analysis of the grassland rotation system see, above all, D. Joravsky, *Ideology and Progress in Crop Rotation* in J. F. Karcz (ed.) *Soviet and East European Agriculture* (1967), pp. 158ff; also, N. Jasny, *Khrushchev's Crop Policy*, (1965), pp. 46ff.
22. Khrushchev, *op. cit.* (in III, 12), p. 348.
23. *Selkhoz*, p. 322.
24. *Ibid.*, p. 48.
25. *Ibid.*, pp. 60f.
26. An unofficial estimate for this period was attempted by J. A. Newth, *Soviet Agriculture: The Private Sector 1950-1959* in *Soviet Studies*, vol. 13, pp. 166ff.
27. *Selkhoz*, pp. 368ff.
28. *Ibid.*, p. 450.
29. N. Nimitz, *Farm Employment in the Soviet Union 1928-1963* in J. F. Karcz, (ed.), *op. cit.* (in VII, 21), pp. 175ff.
30. *Narkhoz 1960*, pp. 742f.
31. Voznesensky, *op. cit.* (in V, 6), pp. 107f.
32. *Narkhoz 1962*, p. 118.
33. Belov, *op. cit.* (in VII, 4), p. 97.
34. *Ibid.*, p. 181.
35. *Ibid.*, p. 102.
36. *Economic Problems of Socialism in the* USSR (1952), p. 103.
37. Belov, *op. cit.* (in VII, 4), pp. 82, 95.
38. *Op. cit.* (VII, 36), pp. 24f.

39. Speech on the 'personality cult', February 1956 in T. Whitney (ed.) *Khrushchev speaks* (1963), p. 258.
40. H. G. Shaffer (ed.), *The Soviet Economy* (1964 ed.), p. 403.
41. *Selkhoz*, p. 41, O. Schiller, *Das Agrarsystem der Sowjetunion* (1960), pp. 51ff.
42. Khrushchev, *op. cit.* (in III, 11), III, 59.
43. Koriagin, *op. cit.* (in IV, 4), pp. 198f.
44. Khrushchev, *op. cit.* (in VII, 39), p. 47.
45. L. O. Richter, *Plans to Urbanize the Countryside* in Degras and Nove (editors), *Soviet Planning* (1964), pp. 38ff.
46. *Op. cit.* (in VII, 36), p. 103.
47. *Ibid.*, p. 100.

CHAPTER VIII

1. In *The Agrarian Question and the Critics of Marx*, chapter 4.
2. *Op. cit.* (in VII, 36), pp. 29f.
3. *Pravda*, March 4, 1951.
4. *Op. cit.* (in VII, 36), pp. 19f.
5. Yefimov, *op. cit.* (in IV, 22), pp. 246ff.
6. Schiller, *op. cit.* (in VII, 41), pp. 30ff. W. A. Douglas Jackson, *The Soviet Approach to the Good Earth: Myth and Reality* in R. D. Laird (ed.), *op. cit.* (in III, 14), pp. 173ff.
7. Yefimov, *op. cit.* (in IV, 22), p. 192.
8. *Narkhoz 1962*, p. 298.
9. *Op. cit.* (in III, 11), vol. I, (speech of January 8, 1955). See also his report of February 23, 1954, *ibid.*, pp. 227-286.
10. *Istoriya Kolkhoznovo Prava* II, pp. 438ff (No. 872). F. A. Durgin Jnr., *The Virgin Lands Programme 1954-1960* in *Soviet Studies*, 13, pp. 255ff.
11. S. I. Ploess, *Conflict and Decision-making in Soviet Russia* (1965), p. 91.
12. *Selkhoz*, p. 223.
13. Speech of December 9, 1963, in *op. cit.* (in III, 11), VIII, 280.
14. Calculated from *Selkhoz*, pp. 636ff.
15. Koriagin, *op. cit.* (in IV, 4), pp. 183, 277.
16. D. Gale Johnson in *Journal of Political Economy* (Chicago, June 1956), LXIV, p. 210.
17. *Op. cit.* (in III, 12), pp. 19f. (Speech of October 17, 1961.)
18. *Ibid.*, p. 333 (March 1962). For good examples of the line of argument used by Khrushchev in the maize campaign, compare *op. cit.* (in III, 11), I, 78ff, II, 80ff.
19. *Narkhoz 1962*, p. 259.

20. e.g. *op. cit.* (in III, 12), p. 414.
21. For a reasoned statement of the official position at the time, see Obolenski (ed.), *op. cit.* (in IV, 4), pp. 90f.
22. M. Y. Lemeshev (ed.), *Ekonomicheskoie obosnovanie struktury selskokhoziaistvennovo proizvodstva* (1965), pp. 12ff.
23. *Narkhoz 1962*, p. 248. Khrushchev, *op. cit.* (in III, 11), IV, 74ff.
24. Koriagin, *op. cit.* (in IV, 4), p. 274.
25. *Op. cit.* (in III, 12), p. 347.
26. *Selkhoz*, pp. 57, 64.
27. *Narkhoz 1962*, p. 330.
28. *Ibid*, p. 342.
29. Calculated from *Narkhoz 1962*, pp. 320, 342 and *Narkhoz 1964*, pp. 390, 400.
30. Yefimov, *op. cit.* (in IV, 22), p. 251.
31. Ploss, *op. cit.* (in VIII, 11), p. 195. For a more realistic appraisal see R. Dumont, *Sovkhoz, Kolkhoz ou le Problematique Communisme* (1964), p. 140.
32. Decree of April 15, 1954; see *Ekonomicheskaya Zhizn* SSSR *1917-1959* (1961).
33. W. W. Eason, *Labour Supply* in Bergson and Kuznets (editors), *Economic Trends in the Soviet Union* (1963), pp. 41ff.
34. *Selkhoz*, p. 450, *Narkhoz 1962*, p. 368, *Narkhoz 1964*, p. 419. (There are minor differences in definition between the figures in *Selkhoz* and later data.)
35. N. Nimitz, *op. cit.* (in VII, 29), pp. 203f.
36. *Op. cit.* (in III, 11), II, 328.
37. Lemeshev (ed.), *op. cit.* (in VIII, 22), pp. 236f.
38. *Selkhoz*, p. 43.
39. Dibb, *op. cit.* (in III, 19), pp. 39f.
40. Koriagin, *op. cit.* (in IV, 4), pp. 62f.
41. *Narkhoz 1964*, p. 377.
42. *Selkhoz*, p. 379, *Narkhoz 1964*, p. 377.
43. *Narkhoz 1964*, pp. 661f.
44. W. Klatt, *The Great Decade of Soviet Agriculture* (1964, mimeographed).
45. Koriagin, *op. cit.* (in IV, 4), p. 193.
46. *Op. cit.* (in III, 12), p. 395.
47. *Ibid.*, p. 396. See Lemeshev (ed.), *op. cit.* (in VIII, 22), p. 180 for the claim that deliveries of tractors and grain combines in 1959-62 were only sufficient to make good wear and tear.
48. *Narkhoz 1962*, p. 298.

49. *Ibid.*, p. 543.
50. *Narkhoz 1964*, p. 659.
51. Koriagin, *op. cit.* (in IV, 4), p. 83.
52. *Op. cit.* (in III, 11), IV, 97f.
53. *Selkhoz*, p. 450.
54. *Narkhoz 1964*, p. 419.
55. Lemeskev (ed.), *op. cit.* (in VIII, 22), p. 267.
56. *Op. cit.* (in III, 11), III, 7.
57. Jasny, *op. cit.* (in VI, 46), p. 422.
58. Lemeshev (ed.), *op. cit.* (in VIII, 22), p. 266.
59. *Ibid.*, pp. 241ff.
60. *Narkhoz 1962*, p. 678.
61. Lemeshev (ed.), *op. cit.* (in VIII, 22), p. 264.
62. M. E. Ruban, *Die Entwicklung des Lebensstandards in der Sowjetunion* (1965), p. 159 (from *Voprosy Ekonomiki* (1959), No. 1, p. 108).
63. *Peasant-Worker Income Relationships* in *Soviet Studies*, vol. 12, p. 21 (1960).
64. Lemeshev (ed.), *op. cit.* (in VIII, 22), p. 163.
65. *Narkhoz 1964*, p. 247.
66. *Op. cit.* (in III, 11), pp. 129f.
67. *Op. cit.* (in III, 12), p. 388.
68. Dienerstein, *Communism and the Russian Peasant*, (1955), p. 83.
69. A. Nove, *Some Thoughts on Soviet Agricultural Administration* in Laird (ed.), *Soviet Agriculture: The Permanent Crisis* (1965), p. 7.
70. Khrushchev, *op. cit.* (in III, 11), III, 46-85, Obolenski (ed.), *op. cit.* (in IV, 4), pp. 213ff.
71. *Op. cit.* (in III, 11), I, 316ff, II, 521ff.
72. *Selkhoz*, pp. 50, 74.
73. *Narkhoz 1964*, p. 562 (footnote).
74. Koriagin, *op. cit.* (in IV, 4), p. 26.
75. *Selkhoz*, pp. 59, 450.
76. Koriagin, *op. cit.* (in IV, 4), p. 310.
77. *Istoria Kolkhoznovo Prava* II, p. 486 (No. 897).
78. *Op. cit.* (in III, 11), V, 436.
79. *Op. cit.* (in III, 12), p. 373.
80. *Ibid.*, pp. 368f.
81. Swearer, *op. cit.* (in III, 14), p. 35.
82. *Narkhoz 1962*, pp. 520, 627
83. *Narkhoz 1964*, pp. 629, 747.
84. *Ibid.*, p. 156.

85. Lemeshev (ed.), *op. cit.* (in VIII, 22), p. 15.
86. Yefimov (ed.), *op. cit.* (in IV, 22), p. 260. The argricultural 'control figures' of the Seven Year Plan are reprinted in Khrushchev, *op. cit.* (in III, 11), III, 321ff.
87. *Ibid.*, p. 194.
88. *Ibid.*, p. 263.
89. *Ibid.*, p. 268.
90. *Ibid.*, p. 269.

CHAPTER IX
1. J. F. Karcz, *The New Soviet Agricultural Programme* in *Soviet Studies*, vol. 17 (1965), pp. 135ff.
2. Koriagin, *op. cit.* (in IV, 4), pp. 78f.
3. Karcz, *op. cit.* (in IX, 1).
4. Koriagin, *op. cit.* (in IV, 4), p. 151.
5. *Narkhoz 1965*, p. 260.
6. *Narkhoz 1964*, p. 254.
7. Koriagin, *op. cit.* (in IV, 4), p. 57.
8. *Pravda*, February 20, 1966.
9. 84,114m. for 1959-1964 ɟ*Narkhoz 1964*, p. 513) *less* 12,978m. in 1959 (*Narkhoz 1962*, p. 434).
10. *Report on the 23rd Congress of the* CPSU (1966), p. 212.
11. e.g. *Molochnaia Promyshlennost*, February and April 1968.
12. *Pravda*, January 25, 1968.

CHAPTER X
1. FAO, *Agricultural Commodity Projections 1975-1985* (1967), vol. I, pp. 71, 144.
2. M. S. Abriutina, *Sel-Khoz. v sisteme balansa nar.-khoziaistva* (1965), p. 102.
3. Koriagin, *op. cit.* (in IV, 4), pp. 286ff.
4. *Narkhoz 1965*, p. 401.
5. *Narkhoz 1962*, p. 327.
6. *Narkhoz 1965*, p. 169.
7. Lemeshev (ed.), *op. cit.* (in VIII, 22), p. 189.
8. *Ibid.*, p. 198.
9. *Ibid.*, p. 111.
10. *Ibid.*, p. 199.
11. Koriagin, *op. cit.* (in IV, 4), p. 148.
12. *Narkhoz 1965*, p. 271.
13. Koriagin, *op. cit.* (in IV, 4), pp. 180ff.
14. *Ibid.*, pp. 279ff.

15. Lemeshev (ed.), *op. cit.* (in VIII, 22), p. 185, Khrushchev, *op. cit.* (in III, 11), IV, 45.
16. *Narkhoz 1965,* p. 401.
17. *Ibid.,* p. 393.
18. Lemeshev (ed.), *op. cit.* (in VIII, 22), pp. 176f.
19. *Ibid.,* p. 188. The United States Department of Agriculture quotes 28,000m. kWh. for 1962 (Stanley G. Brown, US *and Russian Agriculture—A Statistical Comparison,* July 1965).
20. *Narkhoz 1964,* p. 517; Abriutina, *op. cit.* (in X, 2), p. 115.
21. *Narkhoz 1965,* p. 415.
22. Obolenski, *op. cit.* (in IV, 4), pp. 188, 192f.
23. See e.g. N. Jasny, *Production Costs and Prices in Soviet Agriculture* in Karcz (ed.), *op. cit.* (in VII, 21).
24. *L.c.* (in VIII, 16), p. 203.
25. Lemeshev (ed.), *op. cit.* (in VIII, 22), p. 241.
26. *Narkhoz 1964,* p. 396.
27. *Ibid.,* p. 412.
28. *Ibid.,* pp. 254f and calculations therefrom.
29. *Ibid.,* p. 257.
30. V. A. Dobrynin in *Izvestia Timir. sel.-khoz. Akademii* (1966), pp. 199ff.
31. *Narkhoz 1964,* p. 749.
32. Karcz, *op. cit.* (in IX, 1).
33. Dobrynin, *op. cit.* (in X, 30), pp. 206f.
34. Lemeshev (ed.), *op. cit.* (in VIII, 22), p. 240.
35. *Ibid.,* p. 109.
36. *Ibid.,* p. 32.
37. *The Failure of the Soviet Animal Industry* in *Soviet Studies* 15 (1964), p. 291.
38. Lemeshev (ed.), *op. cit.* (in VIII, 22), pp. 37, 171.
39. SSSR *v tsifrakh 1966,* p. 98.
40. Lemeshev (ed.), *op. cit.* (in VIII, 22), p. 97.
41. *Ibid.,* p. 126.
42. *Pravda,* January 26, 1969.
43. Lemeshev (ed.) *op. cit.* (in VIII, 22) p. 140.
44. *Zhivotnovodstvo,* March 1967, p. 48.
45. Lemeshev (ed.), *op. cit.* (in VIII, 22), p. 221.
46. *Ibid.,* p. 248.
47. *Narkhoz 1964,* p. 416.
48. *Narkhoz 1962,* pp. 331, 359.
49. Obolenski (ed.), *op. cit.* (in IV, 4), p. 262.
50. *Ibid.,* p. 366.

51. Lemeshev (ed.), *op. cit.* (in VIII, 22), p. 208.
52. Obolenski (ed.), *op. cit.* (in IV, 4), p. 374.
53. J. W. Mellor, *The Economics of Agricultural Development* (1966), p. 375.
54. *Narkhoz 1965*, p. 420.
55. *Narkhoz 1962*, p. 350.
56. Dumont, *op. cit.* (in VIII, 31), p. 83.
57. Obolenski (ed.), *op. cit.* (in IV, 4), p. 218.
58. *Ibid.*, pp. 64ff.
59. *Ibid.*, p. 76.
60. *Ibid.*, pp. 213ff.
61. A. N. Sakoff, *Production Brigades: Organizational Basis of Farm Work in the* USSR, FAO Monthly Bulletin, January 1968.
62. Obolenski (ed.), *op. cit.* (in IV, 4), pp. 234ff.
63. *Ibid.*, p. 317. According to Schiller, *op. cit.* (in VII, 41), p. 31, a decree of December 1933 fixed the *maximum* arable acreage of grain State farms at 20,000-25,000 ha. and that of their divisions at 2,000-2,500 ha., i.e. well below the optimum dimensions recommended thirty years later by the Agricultural Economics Institute.
64. Obolenski (ed.), *op. cit.* (in IV, 4), pp. 307f.
65. *Ibid.*, p. 273.
66. *Ibid.*, pp. 295ff.
67. *Ibid.*, p. 391.
68. Dumont, *op. cit.* (in VIII, 31), p. 78.
69. *Narkhoz 1964*, p. 411, SSR *v tsifrakh 1966*, pp. 109f.
70. *Narkhoz 1965*, pp. 426f.
71. Obolenski, *op. cit.* (in IV, 4), p. 228.
72. *Ibid.*, p. 239.
73. Dumont, *op. cit.* (in VIII, 31), pp. 238ff.
74. Th. W. Schultz, *Transforming Traditional Agriculture* (1964), p. 113.
75. Mellor, *op. cit.* (in X, 53), p. 375.
76. *Lemeshev* (ed.), op. cit. (in VIII, 22), p. 268.
77. *Ibid.*, p. 163, and *Narkhoz 1962*, p. 232.
78. M. S. Abriutina, *op. cit.* (in X, 2), p. 91.
79. N. Nimitz, *op. cit.* (in VII, 29), pp. 178ff.
80. Abriutina, *op. cit.* (in X, 2), p. 53.
81. Koriagin, *op. cit.* (in IV, 4), pp. 361ff.

CHAPTER XI
  1. Abriutina, *op. cit.* (in X, 2) pp. 80ff.

2. According to *Narkhoz 1965*, p. 168, 25.3 per cent of the wholesale price consisted of turnover tax and 7.1 per cent of profit. As profits were 8,109m. rubles (*ibid.*, p. 759), tax amounted to 28,893m.

3. *Ibid.*, p. 757.

4. Philip Hanson, *The Consumer in Soviet Society* (1968), p. 116.

5. sssr *v tsifrakh, 1966*, p. 132.

6. *Ibid.*, p. 147.

7. *Narkhoz 1965*, p. 408ff.

8. *Narkhoz 1964*, pp. 581ff.

9. As XI, 5, 84.

10. As XI, 5, p. 28.

11. *Selkhoz*, p. 12.

12. As XI, 5, p. 7.

13. *Soviet Agriculture Today*. (Report of the 1963 Agriculture Exchange Delegation.) usda, p. 20.

Index

323

INDEX

Five Year Plan (1928-32), 36, 38f, 49, 51, 97ff, 109, 118, 122f (1933-37) 114, 119 (1946-50) 49 (1966-70) 16, 233ff, 238ff, 248, 254, 256, 296

Flax, 78, 119, 281

Fodder production, 132, 140, 173ff, 179f, 188, 249, 266, 268f, 287, 300

Food, (consumption) 102f, 193, 245ff, 289, 296f (exports) 44, 47, 60, 84f, 94, 120, 192, 248f (imports) 28, 62, 192f (supplies) 21, 28f, 54ff, 68, 102f, 120, 137, 149f, 195, 245ff, 267, 297

Food Processing Industry—see Industry (food)

Foot and Mouth disease, 241, 270

Foreign exchange, 27, 47

Foreign trade, 60ff

France, 61

Fruit (and grapes), 63, 113, 124, 187f, 202, 248, 250, 266, 283, 289f

Fuel, 51, 129, 134, 156, 203

Georgia, 208

Germany, (Eastern) 63 (Nazi) 22, 26, 128, 132ff, 149

Goats, 70, 79, 102, 113f, 133, 141, 163, 179f, 231, 252, 272, 299

Gold, 63

Goods famine, 84, 88

Grain, 13f, 43ff, 51, 55f, 63f, 69, 71, 77ff, 85f, 98, 101, 105f, 112f, 116, 118ff, 132f, 135, 139f, 142f, 150f, 155ff, 159f, 168ff, 179ff, 185, 187ff, 191, 200, 206ff, 214f, 223, 230, 235f, 239, 241, 246, 248, 250, 253, 266, 269, 271, 276, 278, 281, 283, 287, 297 (marketing) 13f, 60f, 69, 73f, 76, 84ff, 93f, 101, 172, 192 (monopoly) 13, 70 (statistics) 40, 118, 189

Grass, 268f

Grassland management system (travopolie), 39, 140f, 161, 175ff

Harvesting period, 126, 253

Hay, 268f

Hides and skins, 60, 72

Horses, 70, 78f, 102, 113, 120, 128, 133, 141, 147, 169, 188

Horticulture, 215

Housing, 58, 195

Hungary, 63

Income (gross and net), 44, 294 (in agriculture) 110f, 121, 129, 183f, 195, 202ff, 214, 222, 224, 236, 256f, 265, 272f, 293f, 297

Indivisible fund, 109f, 255

Industrial inputs, 45ff, 128, 170, 203, 218, 232, 235, 254, 277, 288, 299

Industrialization, 21, 26ff, 34, 36, 42, 52f, 87ff, 121f, 195, 257, 291

Industrial workers, 24f, 68, 120f, 126f, 204f, 291f

Industry, 24ff, 42, 52, 72f, 87f, 128, 152ff, 164, 186, 191, 197, 222, 234, 301 (chemical) 254 (consumer goods) 42, 57ff, 72f, 88, 123f, 137, 152ff, 157, 210, 231 (food) 17, 56ff, 124, 137f, 223, 231f, 240, 267, 294 (heavy) 52, 57, 93, 105, 122f, 152ff (light) 59, 223, 231, 294

Inflation, 120, 136

Infrastructure, 48, 251f, 255, 271, 278

Inter-collective farm unions, 237, 277f, 302

Interest charge, 17, 238

Intervention, 25, 130

Investment, 16, 26ff, 42ff, 59f, 93ff, 101, 109f, 122, 140, 147, 152ff, 164f, 168, 170, 172f, 188, 194ff, 202, 214, 222, 224f, 230, 234f, 250ff, 270f, 273, 293ff, 299, 301 (foreign) 26, 42

Irrigation, 95, 161, 234, 251

Isolation, international, 26ff, 42

Jasny, N., 208, 269

Johnson, D. Gale, 262*

Kamenev, L., 88

Kautsky, K., 50, 93

Kazakhstan, 15, 103, 133, 169ff, 178, 182, 188, 206, 208, 230, 266, 281

Kharkov, 134

Khrushchev, N., 15f, 21f, 32f, 37ff, 108, 119, 131, 141, 156, 159, 161, 163f, 166ff, 182, 186, 188ff, 197, 199f, 204ff, 210ff, 228, 232, 234f, 238, 241, 256, 258, 265, 274, 278, 282, 293, 300

Kolchak, Admiral, A., 69

Kolkhozy—see collective farms

Kosygin, A., 228, 236

Krasnodar, 207, 279

Kronstadt rebellion, 71